On the physical foundations of interstellar space travel

Dr. rer. pol. Erik Kolek

On the physical foundations of interstellar space travel

Chronicles of Business Informatics Physics (CBIP)

Volume 2, edition no. 1.0

2024

Imprint

Kolek, Erik (2024). On the physical foundations of interstellar space travel. In: *Chronicles of Business Informatics Physics (CBIP)*. Volume 2, edition no. 1.0. ISBN: 9783759731159.

Foreword

This book is a pioneering theoretical treatise on Albert Einstein's special and general theory of relativity with the aim of completing and supplementing these theories. As an introduction, the basics of a science of everything are described. This is followed by an extended special theory of relativity and a modern anti-general theory of relativity, both derived by Erik Kolek. The gravitational field of black holes at their event horizon is considered in an outlook, citing the work of Stephen William Hawking. A quantum field theory for the generation and conversion of the heat of light is presented. The work consists of interesting individual contributions on individual topics. This book is a scientific treatise in the field of business informatics on the one hand and physics on the other. The book is aimed at readers with an interest in the subject.

This book is about the physical foundations of interstellar space travel. Interstellar space travel involves traveling between stars, such as between our sun and Proxima Centauri. Humanity, or rather its technologies, are still at the very beginning of the technological development series, and the same applies to the physical foundations. The latter must be listed and explained step by step in order to make traveling between the stars possible, at least in theory.

The problem with physics today is the lack of belief in the possibilities opened up by targeted research into the physical principles of interstellar space travel. For this reason, there are hardly any works that deal with topics such as faster-than-light speed or interstellar travel. The aim of this work is therefore to demonstrate the physical principles of interstellar space travel step by step in such a way that interstellar travel becomes conceivable not only for physicists.

All assumptions in this paper are within the standard model of physics and are therefore based on the state of development of today's science (state-of-the-art). This is a superficial dive into the darkness of the space-time continuum and a brief look behind this standard model of physics is offered. The related, fundamental theories

Kolek, Erik (2024). On the physical foundations of interstellar space travel. In: *Chronicles of Business Informatics Physics (CBIP)*. Volume 2, edition no. 1.0. ISBN: 9783759731159.

that are important for my work can therefore only be found in the original works of Albert Einstein and Stephen William Hawking.

Some of my current research areas and special interests are: (1) cosmological considerations to explain everything and the origin of the space-time continuum, (2) an improved theory of relativity to connect astrophysics with quantum physics, (3) gravity, including quantum gravity, supergravity and megagravity, (4) unified quantum field theories of time, matter, gravity, light, heat, velocity, and (5) quantum technologies.

May 2024.   Dr. rer. pol. Erik Kolek, Diplom-Betriebswirt (FH), M.A., M.Sc.

Kolek, Erik (2024). On the physical foundations of interstellar space travel. In: *Chronicles of Business Informatics Physics (CBIP)*. Volume 2, edition no. 1.0. ISBN: 9783759731159.

Table of contents

Kolek, Erik (2024). On the physical foundations of interstellar space travel. In: *Chronicles of Business Informatics Physics (CBIP)*. Volume 2, edition no. 1.0. ISBN: 9783759731159.

Kolek, Erik (2024). On the physical foundations of interstellar space travel. In: *Chronicles of Business Informatics Physics (CBIP)*. Volume 2, edition no. 1.0. ISBN: 9783759731159.

Kolek, Erik (2024). On the physical foundations of interstellar space travel. In: *Chronicles of Business Informatics Physics (CBIP)*. Volume 2, edition no. 1.0. ISBN: 9783759731159.

Kolek, Erik (2024). On the physical foundations of interstellar space travel. In: *Chronicles of Business Informatics Physics (CBIP)*. Volume 2, edition no. 1.0. ISBN: 9783759731159.

First section: On the foundations of a science of everything

Scientific citation:

Kolek, Erik (2024). A theory of experience via objective inductive reasoning and subjective deductive reasoning leads to the science of everything. In: *On the physical foundations of interstellar space travel.* Chronicles of Business Informatics Physics (CBIP). Volume 2, edition no. 1.0.

Erik Kolek (2024)

A theory of experience via objective inductive reasoning and subjective deductive reasoning leads to the science of everything

Summary

Motivated by the existing conceptual limitations of today's scientific system, which in his experience is based on traditional organizational structures and classical ways of thinking, Erik Kolek, as an expert in theory formation and research methods in business informatics physics, has explored the subjective deductive way of thinking of Albert Einstein and Stephen Hawking in order to solve an existing Copernicus problem based on this, because how else should a single subjectivity deviating from the norm be able to convince a fixed group subjectivity today? This could only succeed if this outdated scientific system completely abandons the inductive objective way of thinking, since, simply put, no objectivity is theoretically conceivable in the universe; only the subjective deductive way of thinking should enable research results that coincide with our universe. Objectivity already falls apart if we generally only think of a circle, because isn't this also a closed curved line? For one person it is a circle and for another it is a closed curved line, which remains independent of an imagined research method, which is supposed to prove objectively inductively which of the two interpretations now reflects the true experiential world? The answer is that both subjectively derived insights are each conceivable as a substitution, which

Kolek, Erik (2024). On the physical foundations of interstellar space travel. In: *Chronicles of Business Informatics Physics (CBIP)*. Volume 2, edition no. 1.0. ISBN: 9783759731159.

corresponds to a both-as-well-as logic that generally appears to be more consistent with the universe than an either-or logic according to Aristotle. Irrespective of the research method, an interpretation of thoughts always remains an imagined interpretation of individuals or groups. However, no new thoughts are generated, not even from intersections, so false truths arise which should be exposed as jointly conceived illusions, especially because often only a textual limited understanding is built up; the visual and mathematical geometric holistic meaning is missing. This work is therefore a theory derived on the basis of personal experience, which, starting from the objective induction way of thinking, describes the obvious advantages of subjective deduction, as this way of thinking leads to a unified science of everything through the fusion of all disciplines, taking practice into account.

A theory of experience via objective inductive reasoning and subjective deductive reasoning leads to the science of everything

In this research article, all statements are based purely on my personal opinion and experience, which I have gained primarily through my private interest in quantum astrophysics and during my time in science in related fields such as business informatics, i.e. generally at universities, colleges and specialist conferences, as well as related observations during my research stays there. Based on this, I would like to form a theory that corresponds to my personal research philosophy, i.e. how I have learned to better understand my more advanced view of science with the help of generally accepted deduction.

In general terms, it could be said (Figure 1) that, due to established, classical organizational structures and cooperative thinking, a way of thinking about induction that only appears to be objective may have prevailed in science. However, this could generally be a false way of thinking, as it would later turn out, because in my opinion the world is very often viewed here as an objective object. The question here is probably not only for me: Can the world as an object be objective at all? The world consists of everything that people are able to imagine in the universe. It is possible for

Kolek, Erik (2024). On the physical foundations of interstellar space travel. In: *Chronicles of Business Informatics Physics (CBIP)*. Volume 2, edition no. 1.0. ISBN: 9783759731159.

me to imagine the world, i.e. the universe, as an object. For example, as a wooden floor embellished with many supposedly different structures and grains, which could remind me or all of us together in general of the planet Jupiter and its well-known red storm eye; more distant wood grains or patterns arising from them could represent space-time for us purely mentally in a model.

So we have learned, according to the theory of experience, that the world should be conceivable as an object, but how do we get objectivity in here or what could represent objectivity without referring subjectively to human thought: Is this philosophy? At this point, it already seems extremely difficult for me to insert an objective thought according to the truth, but I would like to try anyway for my goal of building a single common understanding of science. For example, if we think in a group, we cannot become more objective because it should always be a matter of individual thoughts, the intersection of which should also only represent a subjectivity in the form of agreement; a found agreement therefore does not yet mean a single correct truth. It should be a so-called deceptive, illusionary truth, which in reality, far too often and possibly too late recognized, could be more like a falsehood.

Actually, according to my philosophy of science, I should have shown by now that no objectivity should be found in our thoughts, and if so, it should only appear to be so. Induction means deduction, so can I perhaps gain objectivity in my thoughts through induction based on experience? How much experience do I need for this; is perhaps just a single experience enough? If people in general, and scientists in particular, gain more than one experience, they tend towards an either/or logic according to Aristotle, which means that they try to understand and apply an exclusionary logic because they have learned to do so and it is generally accepted, for example by conducting a literature review on a research question and collecting many aspects (almost like busy bees). Then these appealing (sweet-tasting honey) aspects, which also only represent the thoughts of others or people, are again subjectively attempted to be systematized (for example with thought classification criteria), so that the desired and nevertheless

Kolek, Erik (2024). On the physical foundations of interstellar space travel. In: *Chronicles of Business Informatics Physics (CBIP)*. Volume 2, edition no. 1.0. ISBN: 9783759731159.

deceptive, illusionary objectivity could finally arise. Objectivity still does not seem to be a given to me or to us, which is why I am now trying evidence-based thinking.

Now, however, the aspects to be searched for are specifically related to sustainability and I have found out, for example, with a purely fictitious literature review that there could be social, economic and ecological aspects and I would of course like to try to apply the learned and shared either/or logic in order to remain objective. For example, by conducting a huge, fully comprehensive, i.e. super representative survey of all people in the world and asking them precisely these three aspects of sustainability, which are actually chosen arbitrarily (i.e. subjectively), as one question. As I might have thought beforehand, I then receive the ecological aspects as the answer, but did this really exclude the other aspects or could I have received a generally valid sorting such as ecological aspects, economic aspects and, as the last point, social aspects? So even if everyone in the world were to take part in my online survey as an imaginary research method for generating imaginary objectivity, I would still have received an unimaginably large group opinion, but it should always remain subjective in itself, no matter what research method I come up with.

Objectivity should also be possible in science via a shared way of writing, as I have observed, this is a purely textual, two-dimensional logic or rather Unfortunately, this is usually only a two-dimensional understanding of the text that leads from A to B, B leads to C, C leads to D, but that D is perhaps related to A is not possible because the texts are generally written without a figurative idea; if this is not the case with very few exceptions, then it should automatically be a matter of subjective deductive reasoning, which will be discussed in more detail in the next paragraph. Scientists, like all other people, like to gather in groups, which they call specialist departments, for example, and each specialist department naturally has its own specific tasks, through which they limit their own thoughts, and this could be perceived or seen by an outside observer like me as follows.

Kolek, Erik (2024). On the physical foundations of interstellar space travel. In: *Chronicles of Business Informatics Physics (CBIP)*. Volume 2, edition no. 1.0. ISBN: 9783759731159.

As soon as a new thought is introduced into the established group thinking system, either from within the group or from outside by an individual, the discussion about the new thought begins, i.e. the group tries to objectively check whether the respective thought fits into the already established group thinking system or not. If the idea does not fit, even though it could be true, it is still generally not accepted by the group. This could also be due to the fact that people generally tend to trust other people's observations, they are, so to speak, backed up by the group's decision and therefore believe that they have made an objective decision; unfortunately, this is all too often not the anthropological case according to human intelligence. Thoughts therefore tend to be interpreted individually and in the group in order to find a common denominator, which in turn represents an attempt to create objectivity; the same applies to the example of the literature review as a mostly illusory objective research method.

On the other hand, people also think individually, but often in connection with interpretations of thoughts, which are supposed to merge two different thoughts into one thought, but a thought always remains a thought and should therefore not be interpretable, just as it does not seem to make sense to always look for similarities in the search for truth, because these are also thoughts, here too deceptive, illusionary, i.e. false thoughts are possible, since, generally speaking, the false black swan is hunted mentally; my mental theory swan is therefore always correctly white at the end of the theory formation in every knowledge hunt. An isolated way of thinking through one's own thoughts, such as I only think this or that is right, should also lead to a generally wrong way of thinking, because valuable aspects could be lost here again. The separation of theory and practice therefore seems to make little sense. If, for example, I formulate a theory and neglect established (or at least conceivable) experiences, for example given by the results of experiments already carried out, as in measurement physics, I only get half the truth. This half-truth can also represent a falsity, which in turn could still be perceived (understood) as correctly thought, but should nevertheless ultimately represent an illusion in the mind. Scientists therefore generally also firmly believe in their (published) results, but this belief is also called

Kolek, Erik (2024). On the physical foundations of interstellar space travel. In: *Chronicles of Business Informatics Physics (CBIP)*. Volume 2, edition no. 1.0. ISBN: 9783759731159.

shared knowledge. Many or perhaps even most scientists are anxious about their opinions towards other scientists, just don't say anything wrong, otherwise what might my boss or colleague think of me. Scientists therefore generally tend to act fearfully and are very dependent on the respective group, such as the specialist department in which they find themselves. All in all, objective induction could be a (very often interdependent swarm way of thinking almost like bees or) mostly inefficient group thinking of humans. By now, everyone in the world should really have understood that objectivity should (almost) resemble subjectivity and thus most closely correspond to relativity, which should lead to the elimination of objectivity as such; objective induction therefore seems impossible to me.

In general, however, individual or subjective deduction (i.e. the way of thinking as Albert Einstein or Stephen Hawking did in quantum physics) leads to a specific truth or the only truth. Once this way of thinking has been correctly understood, subjective deduction should no longer lead to falsity, as should be the case with the generally false way of thinking of objective induction in the philosophy of science (Figure 1). This is also one reason why I personally have gotten out of the habit of thinking in terms of objective induction (human intelligence), because it seems to me to be highly inefficient and not an individual way of thinking. Switching from objective induction to subjective deduction is purely a matter of getting used to it and it takes some practice to avoid slipping back into old patterns of behavior (thought patterns).

Of course, I now simply use my imagination, which are invented thoughts on a certain topic, for my thinking and also for the representation of my thoughts in models and visualizations, it supports me very much as soon as I start to think visually, textually and mathematically at the same time; similar to the helpful physics-philosophical case this must have been for Newton in his (subjective) derivation of classical mechanics. Visual logic is always multidimensional and should therefore be superior to purely textual logic, which is mentally two-dimensional without the power of imagination, especially due to the pictorial conception of thoughts, because with such two-

Kolek, Erik (2024). On the physical foundations of interstellar space travel. In: *Chronicles of Business Informatics Physics (CBIP)*. Volume 2, edition no. 1.0. ISBN: 9783759731159.

dimensional texts it almost seems to me as if some content is always missing. Ultimately, this could also be due to the different brains of humans, which could process information differently into knowledge; what is meant here is not the human thought potential as it is generally associated with intelligence, no, I even consider this intelligence potential to be infinitely expandable in us humans (of course there could be exceptions due to biological health conditions), because we may not yet have become fully aware of our thought abilities since human evolution.

Subjective deduction should always lead to individual knowledge that could not be interpreted any deeper or further, such as a circle is round. Could I now think something else here, i.e. interpret it as if it could also be a closed round line? From a textual point of view, apparently yes, but from a visual point of view, no. Here we are back to the point of visual logic with the idea of the content that needs to be experienced, because by means of subjective deduction, everyone should be able to experience the analogy here according to spherical geometry in the mind's eye: round circle equals closed round line. At the same time, this is again a both/and logic, as should be easy to recognize. Such comprehensive knowledge experiences according to the both-as-well-as logic resemble a holistic way of thinking, because information should normally be accumulated until an overall picture of a topic emerges mentally, because if, for example, I now understand the shape of a circle, I could also become aware of the shape of a sphere (without needing further readable information) by imagining a circle as an equator and drawing another circle around it, then a really simple form of spherical geometry should strike me mentally because I could have learned about it.

If a theory is only about thoughts, then it could be really difficult for me not to automatically (almost like a mental algorithm in the brain) connect aspects of experience with regard to practice with it, because, for example, now that I have built up a useful understanding of balls that generally corresponds to nature, I could come up with the idea of letting these two balls roll down an inclined slope purely mentally.

Kolek, Erik (2024). On the physical foundations of interstellar space travel. In: *Chronicles of Business Informatics Physics (CBIP)*. Volume 2, edition no. 1.0. ISBN: 9783759731159.

After all, what would happen if two balls of different sizes each stood on top of a triangle of the same shape and were then released? Everyone should now be able to imagine that one of the balls might reach the bottom faster than the other ball? Would it now also be possible to think about why one ball could reach the bottom faster than the other? To do this, we would have to take a closer look at both balls, because one ball is probably not only bigger, but probably also heavier than the other ball. With this mental observation about a conceivable, i.e. possible property of a ball, we can go back to our mental model, which is based on two different ball sizes that stand on two uniform triangles at the top and should roll down as soon as we mentally release both balls with our fingers. Although this is only a theoretical model, we have now learned for practical purposes that the general rolling behavior of balls could be related to the weight of bodies. But we still don't know why the two balls, although they might be of different weights, because unfortunately we might not have learned how a weight could be specified in uniform scales or with a weight device of the same nature, nevertheless, when we drop both balls, they hit the ground at the same time and why this is so.

What could ball rolling behavior consist of, we might now ask ourselves? To do this, we would have to examine the ball rolling behavior on the triangles more closely. Both seem to move at different speeds, so we could describe this as speed literally, i.e. again mentally, and simply call the difference between the speeds the respective acceleration mentally. So now we should have subjectively deduced a possible (conceivable) result, which we should also need, no more and no less, because why do more work than necessary. We could now make the following statement (axiom): *Heavy spheres generally accelerate more slowly than light spheres and therefore it could also be observed that the movement of heavier spheres seen diagonally downwards in the mind's eye could always simultaneously explain a lower speed than that of lighter spheres, should the general movement of spherical bodies be of interest to the researchers.* So now everyone has learned Galileo Galilei's philosophy of

Kolek, Erik (2024). On the physical foundations of interstellar space travel. In: *Chronicles of Business Informatics Physics (CBIP)*. Volume 2, edition no. 1.0. ISBN: 9783759731159.

science as an example, without having to make any great effort or even memorize the content; that was easy and not difficult, wasn't it?

It is therefore also a matter of theory formation with evaluation, although the evaluation through an experiment, as could now be assumed, could only confirm the imagined contents once again, which should, so to speak, once again make an only apparently objective induction as a generally false way of thinking appear really superfluous. The truth should therefore always be able to produce real results, i.e. results that are usable or that correspond to physical reality, or produce them purely mentally; in this case, knowledge does not mean belief, but knowledge resembles the actual thought (and/or made) experience, since only subjective thoughts about the ball rolling behavior were used for this derivation. Therefore, subjective deduction, although it should never be objective, since everything in the universe should resemble subjectivity, just think of Galileo Galilei's two spheres, could be completely true (i.e. correct) in thought without fear and without having to adhere to a group or its opinion, because this is not only apparently a highly efficient individual way of thinking, which should represent our particular way of thinking as the only true way of thinking of man in our universe. This should also be confirmed if, as a further example, we think of the multidimensional way of thinking introduced with regard to a wooden floor, which could correspond to Jupiter not only purely in our imagination, i.e. with subjective deduction, but also actually in many structures and patterns of physical reality.

Kolek, Erik (2024). On the physical foundations of interstellar space travel. In: *Chronicles of Business Informatics Physics (CBIP)*. Volume 2, edition no. 1.0. ISBN: 9783759731159.

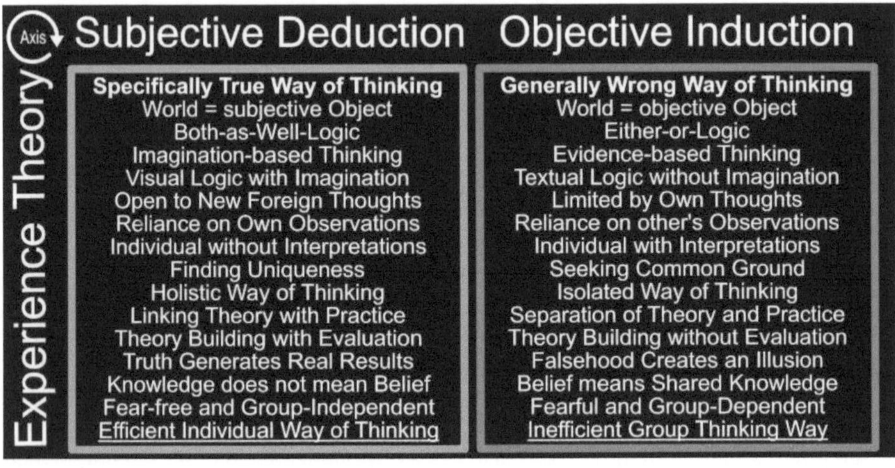

Figure 1. Subjective deduction compared with only apparently objective induction, although both ways of thinking should be similar to the theory of experience.

Since I should have read and understood the works of Albert Einstein that are relevant to me so far, I think, at least for myself, that Albert Einstein could also only have applied a subjective deductive way of thinking or should have followed it in terms of the philosophy of science; as a reminder, this is also my sole personal opinion and experience (Figure 2). In his work, Albert Einstein always accepts only the truth and nothing else, i.e. nothing false, which he ensures by applying his extensive geometric knowledge and thus explaining everything. He therefore examines every possible free thought that he considers relevant to think, and in doing so he is also guided by already established axioms such as those according to Euclid and checks, so to speak, whether the possibility of the thought or its existence is true, i.e. whether the thought can be thought fluently, meaningfully and therefore correctly. Above all, connections with previous knowledge appear to be important, which is why Albert Einstein sometimes repeats what he has previously said. He sometimes uses story-like parables to increase the understanding of physics. He addresses readers, which for me means that he also invites students of other sciences and people with a general interest to follow his thoughts on physics. He uses a vivid pictorial language with many analogies, which

Kolek, Erik (2024). On the physical foundations of interstellar space travel. In: *Chronicles of Business Informatics Physics (CBIP)*. Volume 2, edition no. 1.0. ISBN: 9783759731159.

is why his sentences in principle also represent geometric and/or mathematical meanings to be understood, of course he also uses many formulas for this, but these equations are not pure mathematics, they always describe the universe and the observable behavior of bodies and their interactions, for this he often thinks of coordinate systems.

The key to understanding Albert Einstein's work is to read slowly, picture by picture, imagining, skipping even one sentence without complete understanding, then there is no overall understanding or a faulty understanding. A faulty understanding, for example, is to believe that Albert Einstein describes an area in the sense of a two-dimensional surface like a space-time in which a heavy body like the sun would lie; this is an absolute falsity, because he actually means an at least four-dimensional space-time whose areas can be viewed separately from each other. This is similar to the space-time around the sun, but just like the space-time around the earth; these are different for Albert Einstein and are transformed according to Lorentz, i.e. connected, which could then result in eleven dimensions, so to speak, since three dimensions are required for the connection (transformation) and four dimensions for each event point within space-time, for which only one time dimension can be thought of simultaneously for two four-dimensional space-times according to Albert Einstein. Therefore, only 10 dimensions are conceivable; time does not exist here, but time is relative, which means that it only exists simultaneously between two point events such as between the sun and the earth, but this does not mean that the same time could be set on the sun as on the earth on a clock, because there could also be different time regions due to gravity. That's how I understand it at the moment anyway, which doesn't mean that I couldn't change my opinion or experience again in the course of the next planetary movements.

Albert Einstein only accepts findings from other scientists, mostly physicists and mathematicians, that are conceivable, i.e. true, for him; he normally checks these mentally beforehand and describes them, so that they also appear subjectively

Kolek, Erik (2024). On the physical foundations of interstellar space travel. In: *Chronicles of Business Informatics Physics (CBIP)*. Volume 2, edition no. 1.0. ISBN: 9783759731159.

deductively conceivable to him and can be used; he excludes all others. He always sees and observes his environment, i.e. everything he can see around him. These are the objects, the people, the stars, simply everything, on which he then wants to understand why the behavior he has discovered in objects, people, stars, etc. and how the behavior, i.e. the phenomena that strike him as unusual, could be explained. He is always positive and factual in his way of explaining things (i.e. in his theory of experience according to his reality), always using geometry or mathematics, but always with the idea of physics in the background. He describes his content in real detail and sometimes over several pages, but always without gaps. However, he sometimes asks the reader to think further, that is for him the one who has fully understood his thoughts, so from my point of view it is actually a personal address, especially at a point where everything revolves around the development of a complete theory on the subject of light bodies. His sentence structure is detailed, long, very convoluted, sometimes difficult to read for many people with an inductive objective way of thinking, because the language is now more than 100 years old, but I have become accustomed to this way of writing, so it is a wanting and a being, which is very embellished with anecdotes (short rambling and loosening up stories) and generally a lot of content.

Albert Einstein mentally connects space-time with a rod clock, which means that space resembles a coordinate system on whose axes the various rods (units of length) should be able to be placed differently, he only connects time mentally for us humans with a clock and not because a time exists, he only wants to determine or explain moments of events in time, for which, according to Albert Einstein, two moments always seem necessary. So it is not about time as we live it in everyday life and at work, but about the concept of time, the understanding of time, the definition of time, generally our idea of time (or what we think of time corresponds to time), which appears to be wrong for Albert Einstein, which could be due to the fact that, because of his perceptive powers, he understood the movement of the sun in particular, Perhaps he also perceived the movement of the earth differently with his eyes, as if

Kolek, Erik (2024). On the physical foundations of interstellar space travel. In: *Chronicles of Business Informatics Physics (CBIP)*. Volume 2, edition no. 1.0. ISBN: 9783759731159.

there were twelve hours between morning and evening, i.e. two positions of the sun (places where bodies are located at a certain time) could be clearly determined by a different understanding of time; This understanding of time seems to have been misunderstood not only in physics. If this were not the case, our technological developments would certainly be much more advanced than they are today.

Albert Einstein used evidence, if available, as model components that he incorporated into his theories, such as his (general) theory of relativity. Here it is the objectively non-existent confirmation of the speed of light, which, as far as I know, was only "seen with human eyes" in a liquid on earth in a vacuum and not in the universe, and should therefore only represent an existing limitation of the general law of nature of the constant speed of light c. For him, proofs are also axioms (propositions, expressions, equations) and established general laws of nature, even if they originate from other physicists. In Albert Einstein's work, knowledge arises purely from thoughts, which means that if you understand his thoughts, you have understood his knowledge, which could even go so far that you could absorb his knowledge like a mathematical sequence, which would then be a profound personal change of thought or an increase in human intelligence, and you could almost even form a new inner, mental model, i.e. a new individual reality that could differ from the individual thought realities of other people. Albert Einstein also has critics, despite all the greatness of his theory measured by its degree of truth, I am aware of this, here he tends to meet them with argumentation and in a positive mood, positive means that he generally accepts conceivable, i.e. true knowledge and does not simply ignore (or simply forget) false knowledge, there is no room for group logic here, but he generally values the opinion of other people, because this should simply not seem possible otherwise in the case of Albert Einstein. All in all, Albert Einstein's subjective deduction is probably the most successful way of thinking about physics at the moment for me and certainly for many other scientists.

Kolek, Erik (2024). On the physical foundations of interstellar space travel. In: *Chronicles of Business Informatics Physics (CBIP)*. Volume 2, edition no. 1.0. ISBN: 9783759731159.

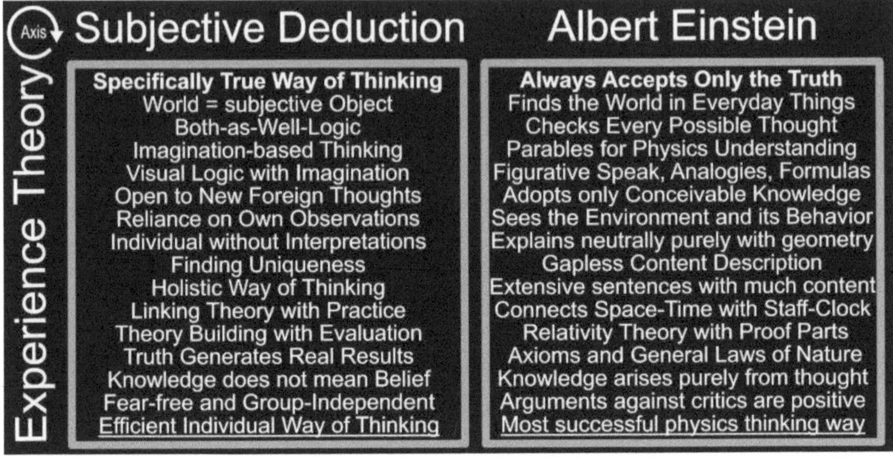

Figure 2. Albert Einstein's subjective deduction is therefore his individual way of thinking according to the theory of experience.

In the meantime, I have already read many of Stephen Hawking's works, especially the specialist articles on his publicly renowned black-body temperature of conceivable space-time singularities, I continue to educate myself with his books in physics, which I should therefore also have understood, I think especially in the case of his dissertation. This means for me, Stephen Hawking could also have used only a subjective deductive way of thinking or should have followed it in terms of philosophy of science; remember it is only my experience and my understanding of everything (Figure 3).

I would use a compound word to describe Stephen Hawking as a brilliant mathematical physicist, because he always accepts only the truth and he is a person who sometimes makes mistakes, really, because he deleted passages in his dissertation, I have seen. I tried to understand why, but overall I would say he wasn't as satisfied with these passages as he was with his other passages that made sense to him both mathematically and physically, because as soon as he could only see mathematical or no physical sense of thought, he crossed out the affected passage in his manuscript. So this was probably the reason. So he finds the world in mathematics

Kolek, Erik (2024). On the physical foundations of interstellar space travel. In: *Chronicles of Business Informatics Physics (CBIP)*. Volume 2, edition no. 1.0. ISBN: 9783759731159.

and uses his mathematical knowledge for physics. I have noticed that Stephen Hawking quite often uses mathematical integration methods, which should be similar to both-as-well-as logic, which should connect different physical facts with each other and therefore represent a mathematical physical evaluation of the respective theory, I noticed this especially in his dissertation with his metrics on the expanding universe, which is really worth taking a look at. In his books, I often find humor hidden between the lines, which reduces the complexity for learners enormously, whereas his technical articles seem extremely stringent and therefore correct (true) as a mathematical consequence of his physics. His physical descriptions are more like equations, so he tends to use a mathematics-based language, because formulas and variables can also contain descriptive physical facts, such as the trend of the expansion of our universe, which can be imagined almost like a living being that is breathing; it grows and then shrinks again, always alternating, but overall our universe grows according to Stephen Hawking's metric.

Stephen Hawking thus shares knowledge that has been verified with logic and thus explains the behavior of our universe; before citing a reference, he certainly always checked whether the respective published result is correct or true. However, he explains logically almost exclusively with mathematics in his specialist articles, here it is a matter of equations that contain his understanding of physics as his reality; Stephen Hawking's physics thought reality differs from that of Albert Einstein, for example; this does not involve different thought opinions, but two different scientific thought styles, both of which follow subjective deduction. It is therefore a more mathematical and physical description of the content of physics. He uses short linked sentences, fewer subordinate clauses, but combined with high precision in the description, I can't find a word that is not in its intended or intended place.

Stephen Hawking generally associates space-time with objects, which means that the universe should represent a thing for him, a kind of object, even time appears to be a thing for him, which is described as a cone starting from a point according to generally

Kolek, Erik (2024). On the physical foundations of interstellar space travel. In: *Chronicles of Business Informatics Physics (CBIP)*. Volume 2, edition no. 1.0. ISBN: 9783759731159.

existing assumptions about light and its speed of movement, he therefore has a different understanding of time or a different thought with regard to time than Albert Einstein, which should be striking not only for physicists. His understanding of time therefore does not seem to me to be wrong, it just mathematically includes Albert Einstein's physical understanding of time or integrates it according to the both-as-well-as logic. The same is generally found in his black body theory, because here Albert Einstein's (general) theory of relativity again plays an important role as part of the evidence for his assumptions regarding the properties of a black hole (space-time singularity) as a conceivable object in our universe. His truth resembles results such as axioms and general laws of nature, knowledge in Stephen Hawking's case only arises from mathematically true results. His understanding of critics seems positive to me, i.e. very open-minded, but here too, mathematical control procedures are certainly used to check these statements about his physics after the intellectual criticism coming from outside. Overall, Stephen Hawking's subjective deduction should represent the most publicly known, generally most accepted and most sympathetic way of thinking about physics, not only in my opinion, but also in the opinion of many other people (scientists) on earth due to his world fame, which Albert Einstein also experienced.

Kolek, Erik (2024). On the physical foundations of interstellar space travel. In: *Chronicles of Business Informatics Physics (CBIP)*. Volume 2, edition no. 1.0. ISBN: 9783759731159.

Experience Theory

Axis | **Subjective Deduction** | **Stephen Hawking**

Specifically True Way of Thinking	Always Accepts Only the Truth
World = subjective Object	Finds the World in Mathematics
Both-as-Well-Logic	Mathematical Integration Methods
Imagination-based Thinking	Humor to Reduce Complexity
Visual Logic with Imagination	Math Language with Descriptions
Open to New Foreign Thoughts	Shares Knowledge Verified with Logic
Reliance on Own Observations	Explains the Behavior of the Universe
Individual without Interpretations	Logically almost only with Mathematics
Finding Uniqueness	Evidential Content Description
Holistic Way of Thinking	Linked Sentences with high Precision
Linking Theory with Practice	Connects Space-Time with Objects
Theory Building with Evaluation	Black Body Theory with Relativity
Truth Generates Real Results	Axioms and General Laws of Nature
Knowledge does not mean Belief	Knowledge Arises only from Results
Fear-free and Group-Independent	Understanding Towards Critics Positive
Efficient Individual Way of Thinking	Most sympathetic physics thinking way

Figure 3. Stephen Hawking's subjective deduction is therefore his individual way of thinking according to the theory of experience.

I have adapted my personal subjective deduction to that of Albert Einstein and Stephen Hawking and try to adhere to it accordingly. I have forgotten or mentally deleted the objective induction generally accepted in science, because it makes today's scientific system seem inconceivable to me due to a lack of objectivity, which I could not even discover or comprehend in a circle, because in a quasi-spherical universe even the smallest circles should be quasi-spherical like spheres (Figure 4).

Just like Albert Einstein and Stephen Hawking, I only ever accept the truth. I find the world as it is in my imagined models, which are textual and mathematically described illustrations of thoughts that are intended to describe the business informatics-physical reality in a way that is as easy to understand as possible. For example, free-flying reference bodies such as a faster-than-light spaceship based on warp quantum technology. I use a systematized assimilation process of insights (knowledge) and technologies (developments) not only with regard to business informatics and physics, in order to really arrive at more advanced quantum technologies and variation stages of these down-to-earth, starting from quantum astrophysically advanced universal foundations, if possible as with a step-by-step construction manual for humanity, easy

Kolek, Erik (2024). On the physical foundations of interstellar space travel. In: *Chronicles of Business Informatics Physics (CBIP)*. Volume 2, edition no. 1.0. ISBN: 9783759731159.

to understand or theoretically designable, by always thinking of the research and practice of space-time body movement (space travel) at the same time.

To support the acceptance of my supersymmetric, relativistic model visualizations (thought mappings), I use hedonistic modelling approaches such as fun and enjoyment of the model through humorous short funny anecdotes, fantasies and ideas. The assimilation process, starting from physics basics up to technology developments, is mentally systematized and thus optimized by an individual collecting algorithm of collective thoughts. This means that I organize suitable thoughts until a conceivable experience results, implemented as knowledge for theory and practice, which should then also appear to be generally true. In my descriptions and explanations, I generally adhere to my multidimensional model language, which, according to its semantics and syntax, adheres exactly to the contents of the model representations in order to simplify understanding. Like Albert Einstein and Stephen Hawking, I only share subjectively derived knowledge, which should clearly emphasize the truth in general as a postulate (principle).

My main aim is to explain the behavior of reference bodies K within reference systems K', K" etc., which coincides with nature, on the basis of their behavior in these coordinate systems. Against this background, I mainly explain geometrically like Albert Einstein, but also often mathematically like Stephen Hawking, if this should appear necessary; these are pictorial, textual model descriptions. I formulate longer content sentences with a lot of statements, subordinate clauses and clever punctuation enable me to integrate additional information to build knowledge and experience.

In my thoughts, space-time resembles a theoretically possible connection, called body-causality, which means this includes all types of bodies, such as the geometrically curvilinear, non-uniform quantum astro-bodies (reference bodies such as particles and planets are to be understood as such) and also an energy field connection between all quantum astro-bodies within a time-space point in the universe that is infinitely small with coordinate systems appears to be subjectively deductively

Kolek, Erik (2024). On the physical foundations of interstellar space travel. In: *Chronicles of Business Informatics Physics (CBIP)*. Volume 2, edition no. 1.0. ISBN: 9783759731159.

conceivable via (probably) electrodynamic quantum gravity as well as loop quantum gravity, which I would like to describe only textually at a later point in this article. I explore the physical foundations and technological developments using a theory of body motion and an associated universal quantum relativity theory as a piece of evidence for theory and practice. This results in axioms, general laws of nature and developments and, above all, the latest highly innovative quantum astrotechnologies for reference bodies and reference systems. For me, knowledge generally arises through the transfer into models and their visualization as individual thoughts. I am open to critics and have a positive attitude, which means that information is examined for falsities after my subjective deductive mental examination and, if available, only stored mentally but not criticized, and I have to realize this in my head, because I only want to publish the truth in all subject areas, including those bordering on business informatics physics. Because my way of thinking should come as close as possible to the two subjective deductive ways of thinking that are generally coincident with science, such as those of Albert Einstein and Stephen Hawking, or is generally based on them. In conclusion, my subjective deductive way of thinking should be one of the currently most advanced universal ways of thinking, which should also appear to be subjectively deductively thought in all existing fields of science today, which, summarized into a single unified theory of science, appear to be more consistent, i.e. to be mentally transferable to the partial theories of the science of components existing according to Aristotle, but best applicable to the unified theory of the science of everything.

Kolek, Erik (2024). On the physical foundations of interstellar space travel. In: *Chronicles of Business Informatics Physics (CBIP)*. Volume 2, edition no. 1.0. ISBN: 9783759731159.

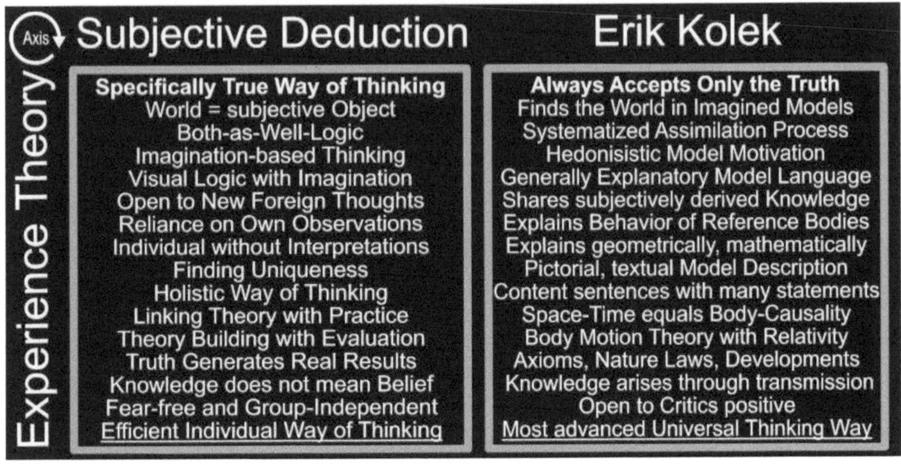

Figure 4. Erik Kolek's subjective deduction is therefore an individual way of thinking according to the theory of experience.

In general, subjective deduction corresponds to a completely different way of thinking than that of only supposedly objective induction, which is still the main focus of today's scientific system, and therefore knowledge should also be acquired much more slowly on the basis of false facts. This means that scientific progress should actually be much faster as soon as the entire scientific system would only deal with subjective deduction (Figure 5). Unfortunately, there are still many barriers at universities, colleges, journals and conferences that should make it difficult to introduce and consolidate a more modern, promising subjective way of thinking. The biggest barrier, apart from classical scientific hierarchies, should be the division of scientific disciplines into individual sub-disciplines, which is probably the biggest and most difficult barrier to overcome. In the interests of the entire scientific system, researchers should initially test out this subjective way of thinking; a before-and-after comparison should be able to resolve any remaining concerns and criticism at least as quickly as the light. No researcher should then want to switch back to classical objective induction as soon as they have become aware of the efficiency advantages in their own thoughts. This is a personal adjustment phase from one way of thinking to the

Kolek, Erik (2024). On the physical foundations of interstellar space travel. In: *Chronicles of Business Informatics Physics (CBIP)*. Volume 2, edition no. 1.0. ISBN: 9783759731159.

other and requires some perseverance and patience, as it is a worthwhile learning process; sometimes people generally revert to old habits of thinking, but it is essential to be aware of this, as it must be prevented mentally. This worthwhile learning process also has nothing to do with high human intelligence; the subjective way of thinking should really just be a different way of thinking that is more in line with the imagination with the greatest possible potential for performance, which is automatically easier for most people to apply through imaginative thinking, without the need for general genius or a special talent. Nevertheless, it could be conceivable that there could be real differences in human intelligence to be experienced, but these should only exist within the respective ways of thinking and not generally in comparison between people thinking with their brains. So something like a neurological distinction between short-term memory and long-term memory might really seem unnecessary with regard to the infinite brain power potential. Not only researchers should try out subjective deduction, but also all other people, because this is the only way to learn for sure, regardless of my theory of experience described here that no objective induction should exist.

Ultimately, today I use subjective deduction in theoretical business informatics physics in order to gain individual insights from this in a mentally constructive way, which I use to answer questions about the tangible world; business informatics physics is a small unified science of everything with the aim of advancing developments of quantum technologies based on physical principles, such as spaceships, their components and the necessary peripherals (Figure 5). The subject areas of this science of everything should be interchangeable, because the more subject areas are used simultaneously, the better the truth about the tangible world should be shaped at the end of the quantum-universal-physical basic developments. Instead of the static triangles, any number of rotatable four-sided pyramids would be conceivable for a large unified science of everything according to the researchers' question, which would be answered by the opposing mental pyramid rotation that they could use as required, For example, the upper pyramid could contain any number of

Kolek, Erik (2024). On the physical foundations of interstellar space travel. In: *Chronicles of Business Informatics Physics (CBIP)*. Volume 2, edition no. 1.0. ISBN: 9783759731159.

interchangeable subject areas such as business informatics, quantum business informatics, quantum field business informatics and relativity business informatics, and the lower pyramid could be labeled physics, quantum physics, quantum field physics and relativity physics. As an example, quantum physics could be visually juxtaposed with business informatics in this scientific theory of everything; any number of triangles or rotating pyramids, each with different subject areas, can be selected and mentally linked to the object of research; in Figure 5, this is the tangible world to which two triangles point.

Anyone who understands this science of everything should now also be able to develop advanced scientific (physical) theories and a down-to-earth warp drive or other quantum technology, as I did. Ultimately, subjective deduction could therefore lead to people possibly being able to utilize their entire brain power potential to infinity, i.e. no longer just trying to interpret experiences of other people in mental sentences (also in the form of texts and images), i.e. the thoughts of other people, in a group-compulsive manner, as before, only experiencing the possible clarity of their minds through this, if they were to deliberately stop these mirage-generating interpretations of thoughts or break the habit through adapted behavior in order to become mentally aware of self-created illusions with a very high proportion of falsity instead of truth, and could therefore also wake up from the daydreams that may constantly exist because of this, because interpreting could be equated with dreaming. Because daydreaming, which probably exists due to the human brain, could really be the reason why not all people today could fully understand the work of Albert Einstein, which is already more than 100 years old, regardless of their intelligence, because this work could not allow any interpretation of thoughts, but would have to be mentally grasped purely with a logical, geometric and not necessarily mathematical understanding; it should actually be sufficient to be able to visualize a circle, because this should enable an understanding of the true, i.e. tangible world, as Albert Einstein shared this physical idea with us (and also Stephen Hawking) to be built up purely

Kolek, Erik (2024). On the physical foundations of interstellar space travel. In: *Chronicles of Business Informatics Physics (CBIP)*. Volume 2, edition no. 1.0. ISBN: 9783759731159.

individually, perhaps with a need for explanation, but without the need for discussion and therefore not dependent on groups.

Finally, I would like to explain what I call the Einstein-Newton exercise, which is a purely mental (theoretical) solution to an n-body problem, so that they can learn the subjective deduction way of thinking more easily through regular model training. All imaginative model ideas are conceivable as training models, which the readers could discover for themselves personally as soon as they understand subjective deduction mentally like a model game and want to adapt its rules. Readers should now visualize the following statements; otherwise, there will be no common understanding or no new thought, i.e. unfortunately only another false interpretation (illusion): We imagine a uniform straight line, on this straight line we mentally place nine circles at approximately the same but nevertheless different distances, we can also imagine these circles as differently colored point events. Now please visualize each point of the circle in different sizes in your mind's eye. The circle on the far left of the straight line is the largest, then comes a much smaller circle, a slightly larger circle, another slightly larger circle, then another smaller circle, followed by the second largest circle on our line, then a slightly smaller circle, then a circle that is more than half the size, then another smaller but comparable circle at the end of the line. Now we have mentally recorded all the necessary circle points and think the line away again so that only the circle points remain. Now we can imagine that the largest circle point is fixed and only appears not to move; we can now let all the other points rotate around this one circle purely mentally. At the starting point of movement of all eight geometric bodies, we will mentally realize that if all eight circle points experience a uniform acceleration of speed, they will each start their movement simultaneously on an almost circular closed curved line, but gradually have to leave their common line position due to the different distances to the largest circle point; in very rare physical cases, however, they should briefly behave again as if positioned on a line to each other. We now imagine all nine circles to be unevenly shaped, they are now elliptically drawn circles and as a next step we leave this two-dimensional model thought fantasy. We

Kolek, Erik (2024). On the physical foundations of interstellar space travel. In: *Chronicles of Business Informatics Physics (CBIP)*. Volume 2, edition no. 1.0. ISBN: 9783759731159.

therefore imagine a three-dimensional world view in terms of geometry; they are now almost egg-shaped spheres (quasi ellipsoids), whereby all other previous thought model fantasies remain intact. To simplify the mental model, we have now omitted certain things such as rings and even smaller spherical ellipses in our imaginary world view; readers can now think of all kinds of detailed thoughts for themselves or subjectively think them away again. A regular repetition of this thought fantasy exercise should specifically promote the subjective deductive way of thinking in the sense of learning, i.e. the experience according to the, in my opinion, unlimited physically existing human intelligence; I thank you for your attention and wish all readers a lot of modeling fun with this Einstein-Newton exercise, which I have subjectively (imaginatively) deduced with regard to an n-body problem in quantum astrophysics (universal physics) and which should also be excellently suited for warming up our brains before performing mental feats.

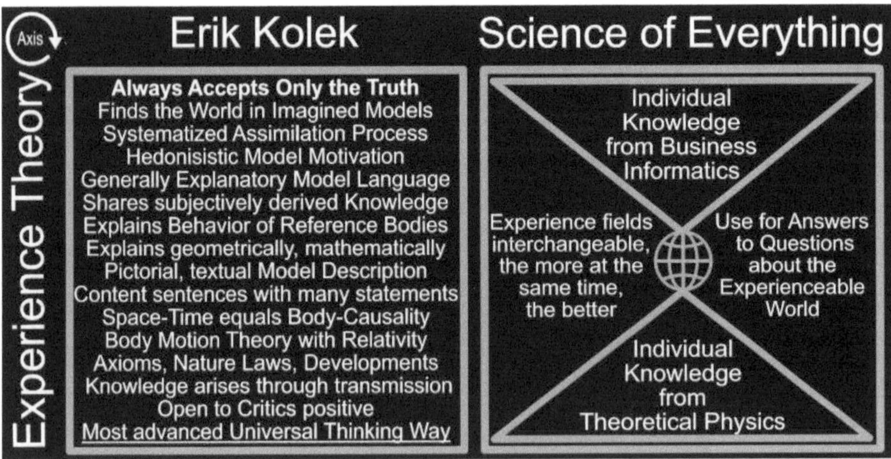

Figure 5. Erik Kolek's subjective deduction is therefore his individual way of thinking according to the theory of experience and is similar to the science of everything as a consequence.

Kolek, Erik (2024). On the physical foundations of interstellar space travel. In: *Chronicles of Business Informatics Physics (CBIP)*. Volume 2, edition no. 1.0. ISBN: 9783759731159.

References

No references have been cited in this research article, as all statements are my personal experience (memory), which I have collected and describe as a comprehensive theory of experience in terms of content.

Table of contents

This research article develops a theory of experience. Various ways of thinking from physics have been introduced. These are intended to renew the content of traditional science. The aim is always to find the truth. Once this has been found, corresponding descriptions must be developed individually.

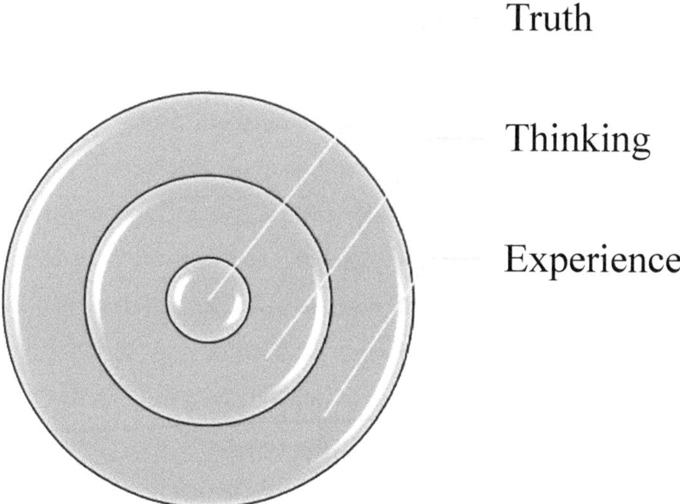

Figure 6. Table of contents illustrated by the importance of the theory of experience.

Kolek, Erik (2024). On the physical foundations of interstellar space travel. In: *Chronicles of Business Informatics Physics (CBIP)*. Volume 2, edition no. 1.0. ISBN: 9783759731159.

Second section: On the foundations of the extended special relativity theory and the anti-general relativity theory

Scientific citation:

Kolek, Erik (2024). Does the inertia of a very small body depend on its energy content?. In: *On the physical foundations of interstellar space travel.* Chronicles of Business Informatics Physics (CBIP). Volume 2, edition no. 1.0.

Erik Kolek (2024)

Does the inertia of a very small body depend on its energy content?

Summary

Motivated by Albert Einstein's special and general theory of relativity, this research article by Erik Kolek assumes a connection between astrophysics and quantum physics, which as a consequence leads to the quantum theory of gravitation. This completely new theory of gravitation should be applicable to very large bodies, but also to very small bodies, as their energy content should have an influence on the inertia of such bodies, for example on black holes and stars as well as on particles and photons. A distinction is also made here between the lightness and darkness of particles and photons. Such a distinction is necessary to design a model-based visualization of a supersymmetric, relativistic universe as it is likely to exist in astrophysics and quantum physics. This model visualization is based on a developed supersymmetric relativistic modeling and visualization approach. This approach was developed by an expert who has research experience in the field of business informatics, therefore this researcher is free of bias and absolutely neutral towards information and knowledge gained regarding theoretical physics. With this method, it is generally possible to conduct research beyond the standard model of physics. The following main results are achieved with the development of the theory: a thermodynamically based further development of Newton's theory of gravitation, the

Kolek, Erik (2024). On the physical foundations of interstellar space travel. In: *Chronicles of Business Informatics Physics (CBIP)*. Volume 2, edition no. 1.0. ISBN: 9783759731159.

discovery of two general laws of nature, known as quantum gravity laws I and II, and, as a further consequence of both laws, the explanation of gravitation in the quantum matter spectrum. Finally, it is discussed why Albert Einstein's special and general theory of relativity is not limited to the speed of light c, because this is physically justified by the fact that it is also suitable for explaining the heat of light Tc, i.e. as a further consequence it represents an advanced molecular-kinetic theory of the heat of light. This completed theory of the heat of light therefore resembles a completely new thermodynamics of physics.

Does the inertia of a very small body depend on its energy content?

The results of a radiation investigation reported by Albert Einstein (1905) in the journal Annalen der Physik led to a very motivating assumption regarding the (special and general) theory of relativity (Einstein, 1905; 1916), which will now be derived (Figure 1). This completed research article refers to original work by Albert Einstein and Stephen William Hawking. To the knowledge of the author, who has research experience in the field of business informatics, these papers represent the only fundamental physical functioning for this new theory of relativity regarding quantum gravity.

According to Einstein (1905, p. 174): "The mass of a body is a measure of its energy content; if the energy changes by L, the mass changes in the same sense [...]".

So when a very large body, such as a star, releases the energy L in the form of mass and radiation, its mass decreases by L/V^2 and the mass of photons (heat of light) is accelerated by $v^2/2$. The kinetic energy of a star is therefore given by equation (1), where V is the speed of light in the form of photons and v is the speed of mass in the form of quantum matter. It is therefore understandable that the kinetic energy K_0 in relation to the system (x', y', z', t') differs from the kinetic energy K_1 in relation to another system (x, y, z, t). This difference of the form $K_0 - K_1$ from equation (1) has an easily understandable physical multidimensional logic. K_0 and K_1 represent the

Kolek, Erik (2024). On the physical foundations of interstellar space travel. In: *Chronicles of Business Informatics Physics (CBIP)*. Volume 2, edition no. 1.0. ISBN: 9783759731159.

kinetic energy values of an identical body, in this case a star, but are related to two coordinate systems that exist in relative motion. A very large body is therefore not in motion in one of the systems (system (x', y', z', t')) and another very large body is in relative motion to this system (system (x, y, z, t)).

(1) $K_0 - K_1 = (+L/V^2 \times v^2/2)$

From an observer's point of view, the mass of photons (heat of light) as seen from the earth, as in the general theory of relativity also published by Albert Einstein (1916), also hits with the speed $v^2/2$. To describe gravitation, the formula of Isaac Newton (1687), which is highly recognized in physics, is applied and extended by the temperature T due to an improved theory of the heat of light, because it is a part of the mass M (2) due to the equivalence principle (Einstein, 1905; 1916), and because of the two masses of the photons (heat of light) and the planet, and because of the distance r between these two. Furthermore, in Newton's theory of gravitation (1687) I added the relative velocity $V_{Rel} = V_0 - V_1$. Therefore, the positive kinetic energy of a star affects the gravity of a very large body like a planet and also of a very small body like a photon when it hits the ground. This first physical fact is given by equation (3).

(2) Force F Newton's theory of gravity improved with temperature $= G \times [(T_1 M_1 T_2 M_2) / (rV_{Rel})^2]$

(3) F Quantum gravity law I = F Gravity $\times$ energy L of the emitting body $= G \times [(T_1 M_1 T_2 M_2) / (rV_{Rel})^2] \times (+L/V^2 \times v^2/2)$

According to the first law of quantum gravity (3), a star emits mass and radiation in the form of heat of light (photons) that hits planets, in this case the Earth, which should influence and increase the gravity on these planets. In other words, the planets are pushed into space-time by the heat of light from the stars. This is due to their own gravity due to their own mass and the additional mass of the photons when they hit the ground of the planets.

Kolek, Erik (2024). On the physical foundations of interstellar space travel. In: *Chronicles of Business Informatics Physics (CBIP)*. Volume 2, edition no. 1.0. ISBN: 9783759731159.

To define the force of quantum gravity (Figure 1), shown here as two laws of quantum gravity (3) and (5), I must first define the negative force or form of energy L of the star that now becomes a black hole. This is because a new theory of gravity should work for both stars and black holes, and preferably for very small bodies.

So when a black hole absorbs the energy L in the form of mass and radiation, its mass increases by L/V^2 and the heat of light (mass of photons) is also accelerated by $v^2/2$. The negative kinetic energy of a black hole is therefore given by equation (4), where V is the speed of the heat of light in the form of photons and v is the speed of the mass in the form of quantum matter. This again makes it clear that the kinetic energy K_0 in relation to the system (x', y', z', t') differs from the kinetic energy K_1 in relation to another system (x, y, z, t). This difference of the form $K_0 + K_1$ from equation (4) has an easily understandable physical multidimensional logic. K_0 and K_1 represent the kinetic energy values of an identical body, in this case a black hole, but refer to two coordinate systems existing in relative motion. A very large body is therefore not in motion in one of the systems (system (x', y', z', t')) and another very large body is in relative motion to this system (system (x, y, z, t)).

(4) $K_0 + K_1 = (-L/V^2 \times v^2/2)$

Again, as seen from Earth, by an observer, as in the general theory of relativity, also published by Albert Einstein (1916), the heat of light (mass of photons) hits the event horizon at the speed $v^2/2$. Therefore, the negative kinetic energy of a black hole affects the gravity of a very large body such as a planet and also of a very small body such as a photon when it hits the event horizon. This second physical fact is given by equation (5).

(5) F Quantum gravity law II = F Gravity $\times$ energy L of the absorbing body = G $\times$ $[(T_1 M_1 T_2 M_2) / (r V_{Rel})^2] \times (-L/V^2 \times v^2/2)$

According to the second law of quantum gravity (5), black holes absorb mass and radiation in the form of heat of light (photons) hitting event horizons, here the Earth

Kolek, Erik (2024). On the physical foundations of interstellar space travel. In: *Chronicles of Business Informatics Physics (CBIP)*. Volume 2, edition no. 1.0. ISBN: 9783759731159.

is not hit by heat of light, which should influence and reduce gravity on these planets. In other words, the planets are being pulled out of space-time by the heat-of-light absorbing black holes. This is due to their own gravity due to their own mass and the lack of mass of the photons when they hit the event horizons of the black holes and no longer due to the planets. So if a black hole arrives near a planet like Earth during the day, it should go dark.

The two laws of quantum gravity (3) and (5) are modelled supersymmetrically, relativistically and visualized to provide an overview and demonstrate their significance for astrophysics and quantum physics. This model-based visualization contains all the physical facts explained above. This model visualization represents a non-flat (!), multi-dimensional region of our expanding universe. It is intended to make the laws of quantum gravity (3) and (5) easy to understand without in-depth mathematical knowledge. Both quantum gravity laws can be applied in astrophysics and quantum physics, as they relate both to the mass and radiation of stars or black holes interacting with planets and to that of light or dark matter particles interacting with light or dark photons. Dark matter particles cannot be hit by photons due to the quantum gravity law I (3). Consequently, photons cannot be hit by dark particles and therefore the quantum spectrum is dark due to Quantum Gravity Law II (5). It therefore appears that an identical physical multidimensional logic applies in quantum physics as in astrophysics.

It may be conceivable to test this quantum gravity theory with very small bodies whose energy content is highly adjustable (for example particles and photons). This quantum gravity theory should still have a bottleneck – the amount of energy that can be produced, but following Einstein (1905, p. 174): "The mass" and temperature of "a" very large and very small "body" are measures of "its energy content; if the energy changes by L, the mass" and temperature "change in the same sense [...]". Thus, following Einstein (1905), I obtain an improved formulation for the energy L, written as the equation $E = TMv^2$ instead of just writing down the well-known equation $E =$

Kolek, Erik (2024). On the physical foundations of interstellar space travel. In: *Chronicles of Business Informatics Physics (CBIP)*. Volume 2, edition no. 1.0. ISBN: 9783759731159.

mc^2. Formulated for several moving masses M, this results in the system of equations $E = \sum(TMv^2)$.

Since this energy formula contains the equivalence principle of Albert Einstein (1905, 1916), the following important model assumption should apply: According to the extended special theory of relativity, it is always conceivable that a black hole could emerge from any star in general, and since the quantum theory of gravitation also includes this possibility of emergence as a consequence of this theory of relativity and is even a model assumption behind the standard model of physics. This star color shift could be the case in particular because once a bright star has ignited, regardless of its size when it ignites, it could become like a dark star as soon as its energy is used up. Therefore, every black hole could in reality be a dark star that has become heavier and whose emitted and absorbed radiation could increasingly take the form of heat instead of light after the color shift. The increased thermal radiation of an ever-darkening (very large) body would mean a significant gravitational increase but would not require gravitational maximization, which could be equivalent to a stellar life cycle related to the light spectrum and, with enough absorbed matter and friction due to heat, could really lead to the re-ignition of dark stars. In this supersymmetric, relativistic model assumption, a dark star would collapse back into a bright star, whereby heavy matter such as iron could condense in space-time due to the rapidly decreasing gravity, which in turn could give rise to new planets. A bright star therefore does not always have to collapse into a black hole due to gravity, because a bright star could just as easily grow into a black hole due to gravity and then collapse back into a white hole. So if bright stars really do grow continuously into dark stars in space-time and can gradually shift their radiation into the infrared light range, then all classes of star colors could also be conceivable, due to the time shift of the star colors in space, starting from violet to green and blue to yellow to orange and red and finally black, as well as according to the time from young to old.

Kolek, Erik (2024). On the physical foundations of interstellar space travel. In: *Chronicles of Business Informatics Physics (CBIP)*. Volume 2, edition no. 1.0. ISBN: 9783759731159.

If this quantum gravity theory is consistent with the physical facts, then mass and radiation transfer inertia between emitting and absorbing very large bodies in astrophysics and very small bodies in quantum physics.

As a brief outlook for the extended special theory of relativity according to the theory of experience, the general assumption of a continuously increasing temperature mass of atoms ($T_{max}\mu_i$), for example, could lead to a more advanced arrangement of the elements according to a stellar life cycle from the formation of a light-bright star to its warm-radiating, probably also radioactive-radiating dark final state (Figure 2). This stellar life cycle could be consistent with the theory of a continuously increasing quantum gravity and thus represent a (new) quantum chemistry, if the physical measurement results in the experiments to determine the atomic masses and the maximum temperatures of the various elements should really agree exactly with the (new) maximizing energy conservation law ($E_{max} = T_{max}M_i$) based on economics (Figure 2).

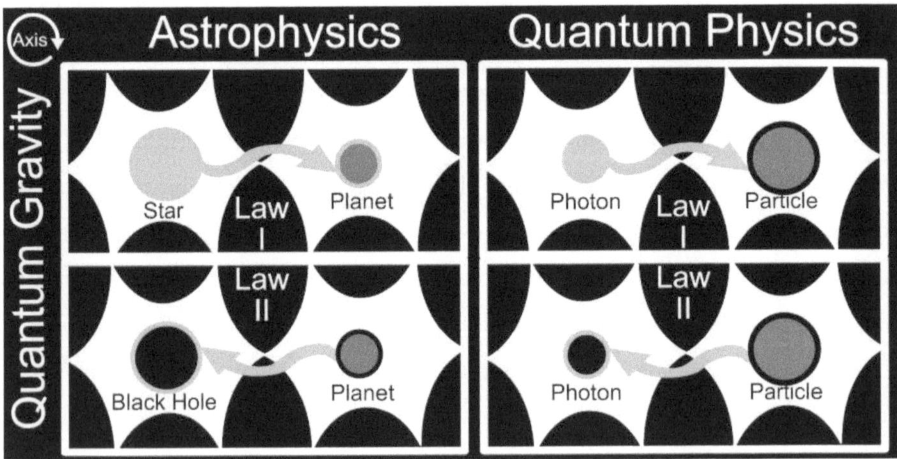

Figure 1. Overview of the two laws of quantum gravity that apply in astrophysics and quantum physics.

Kolek, Erik (2024). On the physical foundations of interstellar space travel. In: *Chronicles of Business Informatics Physics (CBIP)*. Volume 2, edition no. 1.0. ISBN: 9783759731159.

References

Einstein, A. (1905). Ist die Trägheit eines Körpers von seinem Energiegehalt abhängig? *Annalen der Physik 18(13)*, pp. 639–641.

Einstein, A. (1916). Die Grundlage der allgemeinen Relativitätstheorie. *Annalen der Physik 354(7)*, pp. 769–822.

Newton, I. (1687). *Philosophiae Naturalis Principia Mathematica*. 1. edition. Jussu Societatis Regiae ac typis Josephi Streater, London 1687 (http://cudl.lib.cam.ac.uk/view/PR-ADV-B-00039-00001/9 [visited on 08.02.2021]).

Kolek, Erik (2024). On the physical foundations of interstellar space travel. In: *Chronicles of Business Informatics Physics (CBIP)*. Volume 2, edition no. 1.0. ISBN: 9783759731159.

Appendix

1 Xe	2 Rn	3 Kr	4 Ar	5 Ne	6 F	7 O	8 N	9 Cl	10 He
11 H	12 Br	13 P	14 Li	15 S	16 Na	17 Be	18 I	19 Mg	20 K
21 B	22 Se	23 C	24 Rb	25 Zn	26 Ca	27 As	28 Si	29 Al	30 Hg
31 At	32 Cd	33 Cs	34 Mn	35 Sr	36 Te	37 Sc	38 Cr	39 Fr	40 Fe
41 Ti	42 Ni	43 Cu	44 Ga	45 Co	46 V	47 Sb	48 Po	49 Ge	50 Yb
51 Es	52 Eu	53 Fm	54 Ba	55 Ag	56 In	57 Cf	58 Ra	59 Sm	60 Y
61 Tl	62 Sn	63 Bi	64 Tm	65 Pd	66 Pb	67 Rh	68 Ru	69 Zr	70 Pm
71 Dy	72 Nd	73 Ho	74 Nb	75 Pr	76 Mo	77 Er	78 La	79 Tc	80 Ce
81 Tb	82 Gd	83 Au	84 Lu	85 Am	86 Ac	87 Bk	88 Pt	89 Cm	90 Pu
91 Hf	92 Ir	93 Np	94 Pa	95 U	96 Os	97 Ta	98 W	99 Re	100 Th

Legend (cell format):

$$T_{max}\mu_i = n_i$$

Element

Figure 2. Ordering framework of the elements of quantum chemistry sorted according to their rather probable formation according to a generally assumed stellar life cycle with regard to the continuously increasing atomic temperature mass per element ($T_{max}\mu_i = n_i$).

Kolek, Erik (2024). On the physical foundations of interstellar space travel. In: *Chronicles of Business Informatics Physics (CBIP)*. Volume 2, edition no. 1.0. ISBN: 9783759731159.

Table of contents

In this research article, a theory of quantum gravity is developed. This is based on the idea that light consists of light and dark photons (theory of the heat of light). It is convincing that stars and black holes emit and absorb heat of light. This leads to the assumption that the gravity of planets and particles is influenced by emitted or absorbed heat of light.

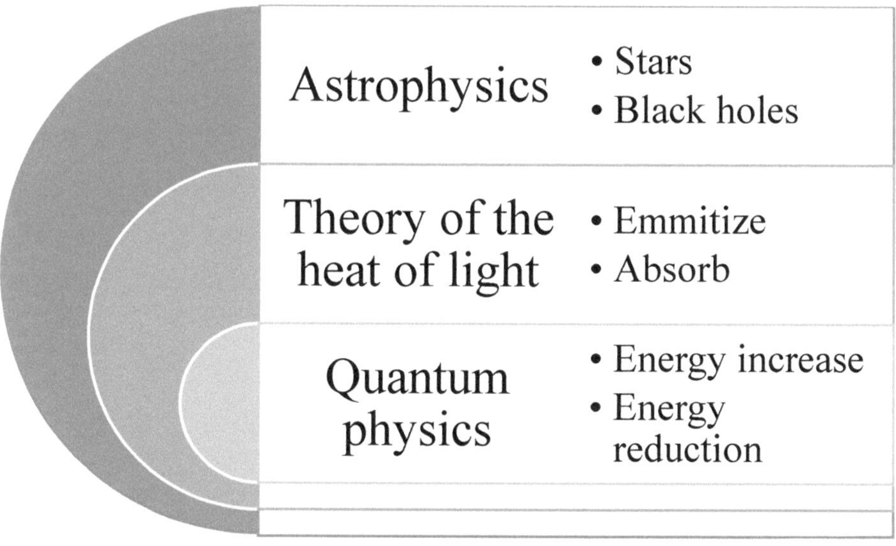

Figure 3. Table of contents illustrated by the importance of the theory of the heat of light.

Kolek, Erik (2024). On the physical foundations of interstellar space travel. In: *Chronicles of Business Informatics Physics (CBIP)*. Volume 2, edition no. 1.0. ISBN: 9783759731159.

Scientific citation:

Kolek, Erik (2024). The basis of the anti-general relativity theory. In: *On the physical foundations of interstellar space travel.* Chronicles of Business Informatics Physics (CBIP). Volume 2, edition no. 1.0.

Erik Kolek (2024)

The basis of the anti-general relativity theory.

Summary.

On the basis of Albert Einstein's general theory of relativity, the basis of Erik Kolek's anti-general theory of relativity is derived, which, like Albert Einstein's general theory of relativity, contains a new gravitational field theory; however, Erik Kolek's anti-general theory of relativity in the physical anti-general case is the anti-gravitational field theory.

The basis of the anti-general relativity theory.

The theory described in this research article forms the most far-reaching extension imaginable of the theory currently generally referred to as the "theory of relativity"; I refer to the latter as the "general theory of relativity" to differentiate it from the former and, like Albert Einstein, also assume it to be familiar (Einstein, 1916). The extension of the theory of relativity was made extremely easy by the design given to the general theory of relativity by Albert Einstein through Minkowski, who was the first to understand the prescriptive equality of the spatial coordinates and the temporal coordinates without error and to make it tangible for the opposite dimensioning of this theory. The mathematical methods necessary for this anti-general theory of relativity were adopted unchanged in the form of the "absolute differential calculus", which was based on the research results of Gauss, Riemann and Christoffel on non-Euclidean manifolds and organized in a system by Ricci and Levi-Civita and already applied to questions of theoretical physics. Like Albert Einstein (1916), I have explained in

Kolek, Erik (2024). On the physical foundations of interstellar space travel. In: *Chronicles of Business Informatics Physics (CBIP)*. Volume 2, edition no. 1.0. ISBN: 9783759731159.

section B of this research article all the mathematical methods necessary for the reader, which cannot be assumed to be familiar, in the most comprehensible and transparent way possible, so that knowledge of mathematical references is unnecessary for an understanding of this presentation. Finally, I would like to take this opportunity to thank my imaginary companion, the physicist Albert Einstein, whose work not only made the study of the relevant physical literature superfluous for me, but also indirectly instructed me as a mentor in the discovery of the field equations of anti-gravity through his sharp-edged logic {effectively}.

A. Fundamental considerations on the relativity principle.

§ 1. Basics of the special relativity theory.

The special theory of relativity is based on the following principle, which is also observed by means of Galileo-Newton's theory: If a reference system K is selected accordingly, so that the laws of physics are valid in their most comprehensible form with respect to it, then *the same* laws of nature are also valid with respect to any other reference system K' that is moved relative to K in a uniform translation. Albert Einstein calls this principle the "special principle of relativity". The term "special" actually means that the principle is limited to one scenario, so that K' has *uniform translational propagation* with respect to K, so that equality with respect to K' and K does not extend to the scenario of *non-uniform* translation with respect to K' with respect to K.

This special theory of relativity therefore does not deviate from Newton's theory due to the principle of relativity, but only due to the principle of the constant speed of light in a vacuum, from which, in conjunction with the special principle of relativity, the simultaneity relativity and the transformation according to Lorentz as well as the laws of nature associated with this with regard to the behavior of propagated practically rigid bodies and clocks follow in a presupposed manner.

Kolek, Erik (2024). On the physical foundations of interstellar space travel. In: *Chronicles of Business Informatics Physics (CBIP)*. Volume 2, edition no. 1.0. ISBN: 9783759731159.

The adaptation that this spacetime theory underwent on the basis of the special theory of relativity thanks to Albert Einstein naturally represents a drastic one; however, the central aspect was still not taken into account. Also according to the special theory of relativity, geometric axioms, for example, are to be understood directly as the laws of position concerning the potential relative locations of (immobile) solids, and (even) more generally the kinematic axioms as axioms that characterize the behaviour of scales (test rods) and clock-hand devices of the same nature. Two marked point masses of a rigid (immovable) body are always equal to a distance of a given number of lengths, without exception, not depending on the position and direction of the body or on the temporal coordinate; two marked clock hand settings of a clock stored relative to the (authorized) coordinate system are always equal to a time distance of a given number of lengths, not depending on the position and temporal coordinate. It soon becomes clear that the anti-general theory of relativity is not (also) able to maintain this understandable physical concept of space-time.

§ 2. The reasons that motivate the extension of the relativity principle.

Newton's theory and also the special theory of relativity have a limitation due to their theory-based findings, which was probably first described in an understandable way by E. Mach. Albert Einstein explains this using the next model example. Two reference bodies {reference masses} consisting of liquid of not unequal geometry and class are freely floating in a space with a correspondingly long distance between them (the same distance also separates them from the remaining point masses), so that only the interactions of gravitation (gravity), which all components of one of the two reference bodies {reference masses} have on each other, have to be taken into account. A distance between the two reference bodies is regarded as a constant{, a straight line linking the two centers of mass always illustrates the same position in the fixed starry sky}. A motion-related relativity with regard to the components of both reference bodies {reference masses} in relation to each other should (still) be neglected. However, all point masses {in total} – viewed by a fixed, immobile observer relative

Kolek, Erik (2024). On the physical foundations of interstellar space travel. In: *Chronicles of Business Informatics Physics (CBIP)*. Volume 2, edition no. 1.0. ISBN: 9783759731159.

to the other point masses – which rotate around an (imaginary) straight line connecting the two point masses consisting of liquid should rotate at a constant angle and constant speed (this is how an imaginable movement according to the relativity of the two point masses is represented). Now Albert Einstein imagined the (countless) surfaces of the two reference bodies {reference masses} (K_1 and K_2), which are determined by means of (relatively fixed) unit rods (test rods) according to measuring physics; it follows that the surface of K_1 above represents a (non-rotating) sphere according to spherical geometry and that of K_2 represents a rotating ellipsoid according to quasi-spherical geometry.

Albert Einstein now questioned this: What reason is there for the reference bodies K_1 and K_2 to differ in their behavior? A satisfactory answer according to epistemology (also called the theory of experience or the theory of experienceability) regarding the last question may only be accepted {defined} if the object conceived as a reason represents a perceptible object of experience (observable with the senses of living beings); since the general law of causality only allows a meaningful assumption (theory, statement) about an experience of the world if the effect and its cause exist inferentially only as perceptible (sensible) objects. Such an answer that is satisfactory in terms of experience theory can of course always continue to be wrong with regard to its physics if it does not agree with the other (acquired) knowledge.

Classical theory cannot provide a satisfactory answer to this question. This includes, for example, the following statement. The laws of motion are valid both for the spatial reference system S_1, with respect to which the reference body K_1 is at rest, but under no circumstances with respect to the second spatial reference system S_2, with respect to which K_2 is at rest. However, the established spatial Galilean reference system S_1, which is thus implemented, corresponds to *a purely invented (or purely fantasized)* cause, (and thus) to no perceptible (sensible) object. This makes it clear that the classical theory does not comply with the condition of the law of causality between cause and effect in this imagined scenario in the (difficult to experience) world of

Kolek, Erik (2024). On the physical foundations of interstellar space travel. In: *Chronicles of Business Informatics Physics (CBIP)*. Volume 2, edition no. 1.0. ISBN: 9783759731159.

reality, but instead only in the (easy to experience) world of reality, in that it recognizes the purely imagined cause S_1 for the sensible different behavior of the reference bodies K_1 and K_2.

A satisfactory answer to the previously opened question may only be formulated accordingly: Due to the (further) reference system formed by means of K_1 plus K_2, which is in accordance with physics, a possible (imaginable) cause is (still) not recognizable {contained} only due to its existence (that is being), on the basis of which the different behaviors of K_1 as well as K_2 could be explained. A (possible) cause should therefore not be found within but outside the reference system (newly formed according to geometry). In general, this leads to the opinion (assumption) [a picture or model that can be imagined according to the theory, such as a metaphorically imagined skewer (distance) on which only one drop of water (reference body) hangs on the left and right] that general laws of mechanics, which specify the design of K_1 and K_2 in detail (specifically), are to be designed accordingly, so that the {usual} motion behavior of K_1 and K_2 must be caused to a large extent by distant point masses, which Albert Einstein (intentionally) did not add to their investigated reference system. Those distant point masses (including their mechanics (movements) according to the relativity with respect to the investigated reference bodies) can therefore be understood as transmitters of fundamentally perceptible (imaginable) causes with respect to the different behaviors of the two considered reference bodies; these have (get) the task (role) of the invented cause S_1. Starting from all conceivable spatial (Galilean) reference systems S_1, S_2, S_i, which are arbitrarily propagated relative to an event point (to each other), none of them can experience a preference by perception in advance (a priori), if the described objection may by no means become relevant again due to the experience-based theory. (General) laws of physics may only be designed accordingly so that they are valid with regard to arbitrarily non-stationary coordinate systems. According to this method, Albert Einstein designed an (experiential or verifiable) extension of his

Kolek, Erik (2024). On the physical foundations of interstellar space travel. In: *Chronicles of Business Informatics Physics (CBIP)*. Volume 2, edition no. 1.0. ISBN: 9783759731159.

principle of relativity; Erik Kolek, of course, did the same in the later sections of this treatise.

In addition to this well-founded view based on empirical theory, however, a (fundamental) truth generally accepted in (theoretical) physics can also speak in favor of an extension of the theory of relativity. K represents a coordinate system according to Galileo, i.e. one which, relative to this system (at least four or more dimensions examined per area or system; no flat area, such as a taut bed in the middle of which a heavy spherical body rests, is meant here), a point mass sufficiently remote from the others propagates uniformly on a straight line. K' represents a second reference system according to Galileo, which is moved relative to K in a uniformly accelerated translation (direction). Thus, relative to K', a point mass sufficiently distant from the others undergoes an accelerating movement in such a way that its increase in velocity and acceleration-translation does not depend on its {or on the natural condition of the considered} elementary connection according to physics or on its physically existing nature.

Is it possible for an observer who is relatively immobile (practically rigid) with respect to K' to formulate a conclusion as to whether he or she is "actually" resting on a coordinate system that is getting faster and faster? The answer to this question is no, since the behavior of separately moving point masses relative to K' just described can just as well be explained using the following approach. The coordinate system K' is not accelerated (i.e. it is at rest); however, a gravitational field exists within this investigated spatiotemporal range, which triggers or causes the acceleration movement of reference bodies relative to K'.

The previous conclusion (theory, assumption) arises as a possibility because Albert Einstein had the experience that he had learned through the existence of the energy field (for example a gravitational field), which has a peculiar character, to assign the same increase in speed to all reference bodies. Eötvös proved in an experiment that this gravitational field has this character together with high precision. The usual

Kolek, Erik (2024). On the physical foundations of interstellar space travel. In: *Chronicles of Business Informatics Physics (CBIP)*. Volume 2, edition no. 1.0. ISBN: 9783759731159.

motion behavior of reference bodies relative to K' is the same as that which, according to experience, appears with respect to reference systems which Albert Einstein usually regarded as "immovable" or "founded" coordinate systems; therefore, this also corresponds to the physically justifiable opinion to conclude that both reference systems K and K' allow an "immovable" view to be found by means of the same observation, or that these are equally suitable as coordinate systems with regard to the representation of processes existing according to physics.

On the basis of these considerations it is generally evident that the establishment of the (anti-)general theory of relativity simultaneously leads to an (anti-)gravitational theory according to experience; since in general an (anti-)gravitational field can be "produced" {"triggered"} by means of a pure change of the reference system. In general, it is also directly recognized (for both theories) that the principle regarding the invariability of the speed of the light body in a vacuum requires modification according to experience. Since it is generally quickly learned that the trajectory of a light body (photon) with respect to K' must generally correspond to a curved trajectory curve as soon as a light body (photon) moves along a straight line with respect to K at a fixed, constant translation speed.

§ 3. The (anti-)space-time continuum. Requirement of (anti-)general covariance with respect to the equations describing the (anti-)general laws of nature.

According to the special theory of relativity and Newton's theory, the spatial and temporal point coordinates in physics have a direct meaning according to logic. A coordinate event has the coordinate x_1 on the X_1 axis, in terms of content this means: A transfer of a coordinate event to this X_1 axis found according to Euclid's theorems of geometry with the help of practically rigid rods remains in place by generally placing a fixed test rod, a uniform scale, (correspondingly often, that is) with an x_1 number on the (non-negative) X_1 axis starting from the starting coordinate of the reference body. A point mass has x_4 equal to t as the X_4 coordinate value, meaning in

Kolek, Erik (2024). On the physical foundations of interstellar space travel. In: *Chronicles of Business Informatics Physics (CBIP)*. Volume 2, edition no. 1.0. ISBN: 9783759731159.

terms of content: A test clock, which is fixed relative to the reference system and (almost) coincides with an event according to its coordinates in space (uniformly constituted), and which ticks (before being set down) according to certain rules, must have passed through x_4 equal to t time intervals (time periods) when a coordinate event coincides (for example, in the case of planet formation by an x_i number of heavy bodies colliding). The determinability of "simultaneity" with regard to event neighbors (point neighbors) directly present in space, or – more precisely formulated – with regard to {the space-time coincidence} of the direct {being side by side} in space-time (point coincidence) was assumed by Albert Einstein, but with a lack of a fundamental term for this, (i.e.) with a lack of a conceptual definition; this definition could, for example, be according to Erik Kolek's opinion: A coincidence of point neighbors according to their coordinates in the relevant system $K(x, y, z, t)$ is a coordinate collision of reference bodies.

Albert Einstein implemented the Galilean coordinate system K with the coordinate numbers x, y, z and t within a (four-dimensional) spatial area that has no gravitational fields, as well as the reference system K' with the coordinates x', y', z' and t', which rotates uniformly relative to K. The common starting point of these two coordinate systems and their Z-coordinate axes should always lie on top of each other. Albert Einstein wanted to learn that for a measurement of the space-time coordinates within the coordinate system K', the previous determination with regard to the physics-related sense of length units and time units cannot be retained under any circumstances. Due to symmetry (coordinate systems lying on top of each other), it is easy to understand that if a circle on the X-Y surface of K around a starting point simultaneously represents a circle on the X'-Y' surface of K'. Albert Einstein now imagined the corresponding circumference and diameter of the circle measured with the aid of a standard test rod (not finitely short relative to the circle radius) and the mathematical result of a division (quotient) of the circumference of the circle divided by the diameter of the circle. In general, if the circle experiment (just imagined) had been carried out using a test rod placed relative to a coordinate system according to

Kolek, Erik (2024). On the physical foundations of interstellar space travel. In: *Chronicles of Business Informatics Physics (CBIP)*. Volume 2, edition no. 1.0. ISBN: 9783759731159.

Galileo K, then the quotient would generally be determined as the numerical value π. In general, this is easy to understand if the entire measurement process is evaluated and observed, starting from the "unmoved" coordinate system K, so that the test rod placed at the edge (on the periphery) receives a reduction according to Lorentz, but the (radial) test rod affecting the radius does not. This is due to the rotational movement of the coordinate system K', whereby for a practically (almost) rigid observer sitting on the circle of K and seeing the circle of K', the rods on the radius appear to be the same length and rods lying on the circumference (edge) of the circle are optically shortened due to the speed of rotation; for an even better understanding, readers can simply think of a circular carousel with two planes. The lower plane, where the first group of readers sits, does not move and the plane above, where the second group of readers sits together with Erik Kolek, does. Therefore, a geometry according to Euclid is invalid with regard to K'; the previously conceived (defined) understanding of coordinates, which does not appear logical without a geometry according to Euclid or whose validity is therefore required, can therefore not be applied with regard to the coordinate system K'. It is also not possible to introduce a general understanding of time within K' that is consistent with the requirements of physics and that can be read with the help of clock hand machines that are fixed (at rest) relative to K' and constructed (created) in the same way. In order to make this comprehensible, readers should imagine one of the two clock hand machines constructed in the same way in the (four-dimensional) axis origin of the coordinate system K(x, y, z, t) as well as at the edge of the circle (circumference, at the periphery of the circle) and look at them starting from the "non-rotating or stationary" coordinate system K. According to one of the (probably) best known results of the special theory of relativity – measured from K (with Euclidean uniform time rods between the clock hand positions) – a clock hand placed on the edge of the circle (circumference) ticks more slowly than the clock hand placed on the starting point, since the first clock hand rotates fixed in K', but the second is not moved at all, but instead lies (practically) rigidly (at rest) in K. An observer sitting (at rest) in the divided origin of both

Kolek, Erik (2024). On the physical foundations of interstellar space travel. In: *Chronicles of Business Informatics Physics (CBIP)*. Volume 2, edition no. 1.0. ISBN: 9783759731159.

coordinate systems, who should also be able to see the clock placed fixed on the circumference of the circle due to light rays, should therefore perceive the clock placed fixed on the periphery of the circle as ticking more slowly (measured between the positions of the hands using Euclidean uniform time rods) than a clock placed fixed near the same observer. Because the observer will under no circumstances be persuaded to decide to bring the (constant) speed of light movement (or time) into dependence (connection) specifically (explicitly) with time (or speed of light) according to the possible visible distances, this observer {the time period} must think his sensory experience further {accordingly} so that a clock hand on the periphery of the circle ticks "in reality" more slowly (more leisurely) than a clock hand placed on the origin of the coordinates. An observer should therefore not be able to allow any neglect, to think (adjust) the understanding of time accordingly, so that the hand speed of a clock is dependent on the position or location.

Albert Einstein therefore obtained the following result: According to the general theory of relativity, it is impossible (really impossible) to define (think about) spatial and temporal units (that correspond to measuring physics) so that the differences (deviations) in spatial coordinates could be measured directly with the (Euclidean) uniform test rod, while time coordinates could be measured with the aid of a generally valid normal clock. Erik Kolek also obtains the same physical fact as the result of the anti-general theory of relativity.

The usual procedure of specifying point coordinates for the space-time continuum in the learned manner is therefore not correct. Moreover, no other method {exclusively the following method} should be suitable either, which could make it possible to align {one of the} reference systems (coordinate systems) with a universe (or a world) with four dimensions, so that through their use a completely easy derivation {design} of natural laws (of a general nature) should be hoped for {effect}; (because this is impossible). The only method consists of considering every conceivable (mental limitations, which resemble an adherence to a fixed connection and its (mental)

Kolek, Erik (2024). On the physical foundations of interstellar space travel. In: *Chronicles of Business Informatics Physics (CBIP)*. Volume 2, edition no. 1.0. ISBN: 9783759731159.

connection with permanence, should not exist here) reference system (coordinate system) in principle equally suitable for the description of (general) laws of nature. This makes the following definition (or principle) conceivable as a condition:

All laws of nature of a general nature must be describable with formulas (equations) which are (equally) valid with regard to each type of reference system (coordinate system), i.e. these must be (generally) covariant (consistent) with possible changes (variables, substitutions).

This should be easy to understand, especially since any physics that resembles this definition (also) resembles Albert Einstein's general principle of relativity (relativity principle). Since all changes (substitutions) must at least also resemble changes (substitutions), (also) every movement relative (with three dimensions) to his coordinate system must resemble (this resembles a conceivable definition of relative movement). This general covariant condition, which deprives space-time of the remaining part of being an object according to physics (objectivity), a condition according to its nature, can be understood by means of the next thought. Every space-time determination (statement, constatement) of humans always resembles the determination of space-time fusion (collision, coincidence). If, for example, the (space-time) event (happening) only consisted of a speed (movement) of mass points (particles of matter), then, by inference, only the collisions (or also approximations, i.e. encounters) of two or more points would have to be observable (or seen). Even the measurement results of humans resemble nothing other than the determination of such meetings (encounters) of mass points (quantum masses) according to the units of measurement (scales, test rulers) of humans with different point masses or coincidences. Coincidences (coincidences of equations) between (different long distances of) clock hands, (different large circles of) clock dials as well as the clock number lines on them (clock dials; points and the arrangement of points on a clock dial) with seen (observed) coordinate events (so-called point events) occurring simultaneously and (at the same moment) in the same place.

Kolek, Erik (2024). On the physical foundations of interstellar space travel. In: *Chronicles of Business Informatics Physics (CBIP)*. Volume 2, edition no. 1.0. ISBN: 9783759731159.

The use of a coordinate system has only one goal, namely to enable a simpler representation (characterization, description) of such coincidences as a whole (i.e. a union of coordinates with respect to a point event). In general, the world (our universe) resembles four spatiotemporal variables (substitutions) x_1 to x_4, so that each event point resembles a reference number system of the variables x_1, x_2, x_3, x_4. Both (or rather all) coinciding (coinciding) event points are equal to the same number system of the variables x_1 to x_4; this means that the coincidence (i.e. the coincidence of the substitutions, variables) is described by a coincidence of the coordinates (variables, substitutions). If, instead of the variables x_1 to x_4, variable (arbitrary) properties (functions) of the same, x'_1 to x'_4 are generally used as an additional reference system so that the coordinate number systems can be exactly connected with each other, then this equation is similar to the axiom (theorem, expression) for the space-time coincidence of both (or more) event points with regard to each of the four coordinate variables also in the second coordinate system. Since all experiences (memories, thoughts) of humans can be explained by corresponding coincidences (correspondences) according to physics, there seems to be no justification (mentally) for determining or preferring one reference system (coordinate system) over another as more suitable, i.e. Albert Einstein arrived at the condition of general agreement; (the covariance of equations with regard to space-time demanded by Albert Einstein is addressed here). This demand for a general covariance must also remain unchanged as a basis of the anti-general theory of relativity, since Albert Einstein's general principle of relativity is still valid here.

Kolek, Erik (2024). On the physical foundations of interstellar space travel. In: *Chronicles of Business Informatics Physics (CBIP)*. Volume 2, edition no. 1.0. ISBN: 9783759731159.

§ 4. {A basic measurement function} Connection of point coordinates on four dimensions {for space as well as time} with regard to space-time with measurement experiences (measurement results).

The equation on which the investigation is based (as a meaningful theorem or an analyzing expression, i.e. an axiom) with regard to the gravitational field.

It was not important to Albert Einstein in his (subjective) derivation (deduction) to describe his general {or the design of his} theory of relativity as a system of logic (subjective deductive way of thinking) that was extremely easy to understand with as few (integrated, thought) propositions (equations, expressions, axioms) as possible. Instead, Albert Einstein's main aim was to formulate {show} his general theory of relativity directly so that an observer understands the mental (i.e. psychologically existing) nature of his applied method and that his presupposed foundations are hopefully comprehensible with the help of the theory of experience. Albert Einstein now introduced this basis in accordance with this meaning or sense:

With respect to never-ending (infinite) tiny space-time regions on four dimensions, (therefore no two-dimensional regions or surfaces can be thought of here), Albert Einstein's (general including the special) theory of relativity appears to be (really) true in a smaller (narrower) sense for the physical case of a coincident selection of coordinates.

The velocity of a movement (acceleration state) of a coincident infinitely tiny ("spatial") reference system (coordinate system) must be selected in a compensating (corresponding) manner so that no gravitational field can arise; this appears to be conceivable for an infinitely tiny space-time range. The space coordinates should be given by {were named by Albert Einstein as} X_1 to X_3; X_4 equals the assigned temporal coordinate, (which is also a variable), which is measured with a suitable scale. {A time period or the time scale (time scale, time setting), which is measured

Kolek, Erik (2024). On the physical foundations of interstellar space travel. In: *Chronicles of Business Informatics Physics (CBIP)*. Volume 2, edition no. 1.0. ISBN: 9783759731159.

with "light units", must be selected so that the speed of light movement v {vacuum (light) speed} – measured within the "spatial" reference system (coordinate system) – can be set equal to 1.

All four coordinates together or Coordinate systems have a direct physical sense (1a) according to the special theory of relativity if they are imagined as an invariable (rigid) rod, i.e. as a predetermined uniform scale (test rod), in the case of a predetermined direction of the reference system; (in the case of two coordinate systems conceived as two point events, this description is given by $dsds = ds^2$ and the negative sign describes an infinitely small space, i.e. the idea of an unimaginably small size in which, however, simultaneity still exists, which is why the physical sense is given a positive sign here).

(1a) $+ ds^2 = - dX_1^2 - dX_2^2 - dX_3^2 + dX_4^2$ (Einstein, 1916, p. 777).

(1b Parallel universe) $- ds^2 = + dX_1^2 + dX_2^2 + dX_3^2 - dX_4^2$ (Einstein, 1916, p. 777).

Equation (1a) therefore has a measure coincident with the special theory of relativity that is not dependent on the direction of the spatial reference system and can be determined or experienced with a spatial time count. Albert Einstein describes ds as belonging to the rod (or unit) conceived with respect to the endless (small) point neighbors within the space with four dimensions coincident with the corresponding reference line element; (this is a coordinate system that appears conceived on the smallest local dimension like such a mini line element) {, this appears as a value obtained by means of rod ablation}. If the quantity (ds^2) assigned to this line element $[dX_1$ to $dX_4]$ has a positive sign, then Albert Einstein, on the basis of Minkowski, described it as a time-determined element [reference system with dX_1 to dX_4] equal to its quantity determined by time (ds^2), and conversely, with a negative sign, as a space-determined element (reference system with dX_1 to dX_4) equal to its quantity determined by space (ds^2).

Kolek, Erik (2024). On the physical foundations of interstellar space travel. In: *Chronicles of Business Informatics Physics (CBIP)*. Volume 2, edition no. 1.0. ISBN: 9783759731159.

With regard to the examined "reference line element" or with regard to the two event points located endlessly next to each other, fixed increments (differentials, summations) dx_1 to dx_4 are also assigned to the point coordinates with four dimensions of the selected line element system. If the line element just described and a "spatially" determined reference system exist or are stored with regard to the point under investigation, then the dX_γ must be described at this point by means of a defined rectilinear uniform function of the dx_σ:

(2) $dX_\gamma = \sum_\sigma(\alpha_{\gamma\sigma} \times dx_\sigma)$; with one unknown alpha for two line elements, however, is considered and summed starting from σ (Einstein, 1916, p. 778).

If the statements of equation (2) are inserted into the formulations of equation (1a), the following equation (3) generally results:

(3) $ds^2 = \sum_{\sigma\tau}(g_{\sigma\tau} \times dx_\sigma \times dx_\tau)$; with a variable for gravity derived from two summed line elements $\sigma\tau$ (Einstein, 1916, p. 778).

The mathematical constructs $g_{\sigma\tau}$ resemble the properties of x_σ, which must now have lost their dependence with regard to the direction as well as the state of motion of the "spatial" reference system; since ds^2 resembles a magnitude value that can be determined by means of space-time measurement with scales and clocks {space-time counting}, assigned to time-space-related, endlessly adjacent event points (analogous to the Big Bang) and conceived without dependence with regard to all arbitrarily selected coordinates. The constructs $g_{\sigma\tau}$ must be selected accordingly so that $g_{\sigma\tau}$ equals $g_{\tau\sigma}$; a summation must include all quantities with respect to τ and σ {or the link} so that a summation can be formed using 4 times 4 variables (constructs) to be added, with respect to which 4 pairs must be unequal.

According to Albert Einstein's standardized theory of relativity, a consideration with the help of the subject under investigation here (gravitation) {by means of anti-generalization} becomes possible {meaningful} if it {allows}, due to the special behavior of all $g_{\sigma\tau}$ within a non-infinite time-space range, to select a coordinate system

Kolek, Erik (2024). On the physical foundations of interstellar space travel. In: *Chronicles of Business Informatics Physics (CBIP)*. Volume 2, edition no. 1.0. ISBN: 9783759731159.

within this range in a coincident manner so that all $g_{\sigma\tau}$ correspond (mentally) to the {in the special case} constant quantities (4a). Albert Einstein later showed that a selection of coincident coordinate quantities must generally be excluded with regard to non-infinite ranges.

$$\text{(4a)} \quad \begin{vmatrix} -1 & 0 & 0 & 0 \\ 0 & -1 & 0 & 0 \\ 0 & 0 & -1 & 0 \\ 0 & 0 & 0 & +1 \end{vmatrix} \quad \text{(Einstein, 1916, p. 778)}.$$

$$\text{(4b Parallel universe)} \quad \begin{vmatrix} +1 & 0 & 0 & 0 \\ 0 & +1 & 0 & 0 \\ 0 & 0 & +1 & 0 \\ 0 & 0 & 0 & -1 \end{vmatrix} \quad \text{(Einstein, 1916, p. 778)}.$$

Based on the investigations contained in sections 2 and 3, it is thus learned that the functions $g_{\sigma\tau}$ are analogous to properties that the gravitational field (multidimensional field consisting of gravity) specifies with regard to the selected coordinate system. As an example, Albert Einstein initially wanted to assume that his special theory of relativity applies to a coincident (and therefore only suitable) selection of (four) coordinates and that it is therefore possible to apply (practice) to any {sub-area} to be investigated with four dimensions. The properties (constructs) $g_{\sigma\tau}$ thus receive the numbers assumed in the vector (4a). Every free-flying mass point is therefore accelerated linearly and homogeneously with respect to this vector system (i.e. with a straight-line uniform velocity v). If an arbitrary change is now generally completed, which leads to the renewed (changed) space-time coordinate values x_1 to x_4, then in this changed (renewed) space-time system the functions $g_{\sigma\tau}$ do not represent invariant {space-time properties} as before, instead they are (changeable) space-time properties. Therefore, every directional movement of a separately existing (single, not bound, free) matter particle (matter point, mass quantum) must be observed (seen)

Kolek, Erik (2024). On the physical foundations of interstellar space travel. In: *Chronicles of Business Informatics Physics (CBIP)*. Volume 2, edition no. 1.0. ISBN: 9783759731159.

simultaneously within the changed (renewed) coordinate values analogous to a non-linear (curved, crooked), heterogeneous (non-uniform) locomotion (propagation). However, with regard to the flying point mass or its nature, this law of mechanics (i.e. Albert Einstein's general theory of relativity and my counter-theory to it) exists without a dependency. Albert Einstein understood this directional movement to be analogous to a movement that arises due to the influence of a gravitational field. Albert Einstein understood that the existence of a gravitational field is associated with a space-time substitution (space-time change) of the space-time properties $g_{\sigma\tau}$. Also in this general consideration, Albert Einstein had to maintain this derivation (treatise, opinion, way of thinking) that he was not able to establish a significance (relevance) of his special theory of relativity within a non-infinite range with a coincident selection of coordinates. This then led to the gravitational field being characterized by the space-time properties $g_{\sigma\tau}$.

According to Albert Einstein's general theory of relativity, gravitation (gravity) performs a special task (or is different) with regard to all other interactions, especially with regard to all interactions due to electromagnetism, because the 10 properties $g_{\sigma\tau}$ that determine the gravitational field also specify the measurable space (according to physics) with four dimensions based on its metric (space-time) properties. This is to be understood in the same way with my anti-general theory of relativity, because, for example, one meter on the earth is not equal to one meter on the moon. Albert Einstein and I also have 10 but not 11 dimensions $g_{\sigma\tau}$, because two line element systems (2 × 3 spatial coordinates) are considered simultaneously (2 × 1 time coordinate). As in the special theory of relativity, a connection is made here via a three-dimensional space-time transformation (space-time substitution) (1 × 3 space coordinates); accordingly, theories according to general relativity with more than 10 dimensions $g_{\sigma\tau}$ correspond to a falsity (or a fallacy).

Kolek, Erik (2024). On the physical foundations of interstellar space travel. In: *Chronicles of Business Informatics Physics (CBIP)*. Volume 2, edition no. 1.0. ISBN: 9783759731159.

B. Basic ideas regarding the mathematics of the derivation of general covariant equations.

Now that Albert Einstein learned in part A that the general principle of relativity points to a condition that in physics all {generalized} reference systems (coordinate systems, systematized {expressions}) may (only) be covariant with respect to arbitrary (all) changes (transformations, substitutions) of the coordinate quantities x_1 to x_4, Albert Einstein had to think about the method by which such generally covariant systems of equations are to be derived. Albert Einstein now learned this completely mathematical method; whereby he learned that the assumed (for the most part non-variant, i.e. a barely variable constant) invariant *ds* has an important task during the application of the method (solution method) with regard to the system of equations (3), which Albert Einstein imagined to be analogous to a "line element system" (or coordinate system in his mind) according to Gaussian surface theory.

{§ 5. A line-like geodesic (equation of motion for coordinate systems).}

The underlying idea here as a fundamental idea is similar to the following according to Albert Einstein's general theory with regard to covariants. A certain construct (object, thing) is conceived as a "tensor" ("model") with respect to all reference systems (line element systems) as a set of spatial properties, each of which is designated as a "component" (or "constituent") of a tensor (model). For this purpose, there are certain guidelines according to which the tensor components for a modified (renewed) line element system can be calculated if they exist as given with respect to the previous line element and if the substitution (transformation) that connects the two line systems (also) exists as given. Each construct (i.e. each object in the model) named as a tensor (model) {for the special case} must also be described {designed} in such a way that the transformation systems (equation systems) exist in a straight line and uniformly in relation to their components (tensor components, model components). In conclusion, all components must dissolve in the modified (renewed) line system, provided that they (also) dissolve completely in the previous element

Kolek, Erik (2024). On the physical foundations of interstellar space travel. In: *Chronicles of Business Informatics Physics (CBIP)*. Volume 2, edition no. 1.0. ISBN: 9783759731159.

system (dimension system). If a law of nature is formulated by zeroing {resolving} each element (or component) of a model (or tensor), then this becomes a generally covariant law of nature; by analyzing the design laws (modelling conventions, design science) of the models (tensors), Albert Einstein (now) understood the basics regarding the derivation of generally covariant equations.

§ 5. The contravariant and covariant four-vector.

The contravariant four-model vector. The line element system (line construct model) {with four dimensions} is thought to exist by means of the four {"projection parts"} "components" dx_v, whose transformation method is formulated by means of the system of equations (5).

(5) $dx_\sigma{}' = \sum_v[(\Omega x_\sigma{}'/\Omega x_v) \times dx_v]$ (Einstein, 1916, p. 780)

All functions $dx_\sigma{}'$ are rectilinear and uniformly determined by means of the dx_v; Albert Einstein therefore (also) called such infinitely small quantities (differentials) of the coordinates dx_v constituents (components) of a "model" (tensor), which he called a *contravariant four-vector* in a special sense. All model constructs (model objects, model things, model contents) that exist with respect to the reference system (line system) by means of the four values A^v, which are determined according to the same transformation method (equation method), were also called (in the same special sense) *contravariant four-vector* by Albert Einstein.

(5a) $A^{\sigma'} = \sum_v[(\Omega x_\sigma{}'/\Omega x_v) \times A^v]$ (Einstein, 1916, p. 780)

The transformation method (5a) simultaneously produces the sequence, so that the (solutions) summations $[A^\sigma \pm B^\sigma]$ also resemble the components of a four-vector, if A^σ and B^σ are components. The same is true with regard to any system subsequently incorporated as a "model" (tensor) (this is the addition-subtraction method with regard to models (tensors)).

Kolek, Erik (2024). On the physical foundations of interstellar space travel. In: *Chronicles of Business Informatics Physics (CBIP)*. Volume 2, edition no. 1.0. ISBN: 9783759731159.

The covariant four-model vector. The four values A_v are equal to the components of a covariant four-vector if B^v applies to any arbitrary selection of a contravariant four-model vector (, here the annotation v is at the top to indicate the class of the four-vector).

(6) $\sum_v(A_v B^v) = Skalar\ (ds)$ (Einstein, 1916, p. 781).

Equation (6) can be used to determine the transformation method with regard to a covariant four-model vector as a sequence {directly}. In general, for example, in the system of equations

$\sum_\sigma(A_\sigma{}'B^\sigma{}') = \sum_v(A_v B^v)$ (Einstein, 1916, p. 781)

B^v in the right-hand term is replaced by the lower formulation, which is created using the inverse of the system of equations (5a),

$\sum_\sigma[(\Omega x_v/\Omega x_\sigma{}') \times B^\sigma{}']$ (Einstein, 1916, p. 781)

then generally results in

$\sum_\sigma(B^\sigma{}')\sum_v[(\Omega x_v/\Omega x_\sigma{}') \times A_v] = \sum_\sigma(B^\sigma{}'A_\sigma{}')$ (Einstein, 1916, p. 781).

This results in a further sequence, since within this system of equations one $B^{\sigma\prime}$ in each case is not mutually dependent, i.e. can be determined individually (and thus shortened), the transformation method

(7) $A_\sigma{}' = \sum[(\Omega x_v/\Omega x_\sigma{}') \times A_v]$ (Einstein, 1916, p. 781)

Side note for facilitating the writing method of equations.

If you look at the systems of equations in this section, you will learn that the data (indices) that are annotated more than once to a summation symbol [for example, the data v within (5)] must always be added up, but only with the help of data that appears twice. Consequently, it is not impossible to delete all summation symbols without affecting understanding. Albert Einstein used the following rule for this purpose: If a

Kolek, Erik (2024). On the physical foundations of interstellar space travel. In: *Chronicles of Business Informatics Physics (CBIP)*. Volume 2, edition no. 1.0. ISBN: 9783759731159.

statement occurs twice within a term of an equation, it must always be added with regard to this index if there is no contradiction in the equation.

The transformation method [(5) or (7)] establishes the inequality (difference) with respect to the contravariant and covariant four-vector. These two methods resemble models (tensors) according to the previous general marginal note; that is their purpose. Based on Ricci and Levi-Civita, the covariant four-vector must be characterized by a lower index, the contravariant four-vector by an upper index.

§ 6. Tensors with two and larger ranks.

The contravariant model tensor. If Albert Einstein determined all 16 multiplication results (products) $A^{\mu\nu}$ of the components A^{μ} and B^{ν} of two contravariant four-vectors

(8) $A^{\mu\nu} = A^{\mu}B^{\nu}$ (Einstein, 1916, p. 782),

then $A^{\mu\nu}$ according to (5a) and (8) corresponds to the transformation method

(9) $A^{\sigma\tau\,'} = (\Omega x_\sigma\,'/\Omega x_\mu) \times (\Omega x_\tau\,'/\Omega x_\nu) \times A^{\mu\nu}$ (Einstein, 1916, p. 782).

Albert Einstein called {16 values as} a construct which is characterized with respect to all coordinate systems by means of 16 values (properties) which comply with the transformation method (9), a contravariant model tensor with rank 2. By no means every such model can be modeled according to (8) by means of two four-vectors. However, it seems {still} easy to understand that the 16 arbitrarily determined $A^{\mu\nu}$ can be modeled as the summation of $A^{\mu}B^{\nu}$ given by four coincident pairs of four-vectors. Therefore, in general, almost any of the expressions which are valid with respect to the imaginary model tensor of rank 2 are most easily shown by proving them in general with respect to special model tensors of the kind (8).

The contravariant model tensor with arbitrary rank. It is easy to understand that, according to (8) and (9), contravariant model tensors with a rank of 3 and above are also conceivable by means of 4 to the power of 3 etc. components. It is also to be understood with the help of (8) and (9) that in general, according to the same sense, the contravariant four-vector is to be understood analogously to a contravariant model tensor with rank 1.

The covariant model tensor. If the 16 multiplication results (products) $A_{\mu\nu}$ of the components of two *covariant* four-model vectors A_μ and B_ν are generally determined in opposite directions

(10) $A_{\mu\nu} = A_\mu B_\nu$ (Einstein, 1916, p. 782),

then the transformation method is valid with respect to these covariant four-vectors

(11) $A_{\sigma\tau}{'} = (\Omega x_\mu/\Omega x_\sigma{'}) \times (\Omega x_\nu/\Omega x_\tau{'}) \times A_{\mu\nu}$ (Einstein, 1916, p. 782).

With the help of this transformation method, the covariant model tensor with rank 2 is conceived. All thoughts regarding the contravariant model tensor are also valid for the covariant model tensor.

Side note. To aid understanding, the invariant (also known as a scalar) can be equated with a contravariant model tensor with rank 0 or a covariant model tensor with rank 0.

The mixed model tensor. It is also (generally) possible to create a model tensor with rank 2 of the type

(12) $A_\mu{}^\nu = A_\mu B^\nu$ (Einstein, 1916, p. 783),

which is covariant with respect to the specification μ and contravariant with respect to the specification ν. The transformation method is

(13) $A_\sigma{}^\tau{'} = (\Omega x_\tau{'}/\Omega x_\beta) \times (\Omega x_\alpha/\Omega x_\sigma{'}) \times A_\alpha{}^\beta$ (Einstein, 1916, p. 783).

There certainly exist mixed model tensors with contravariant and covariant expressions, each with an infinite number of specifications (indices). (Mixed) covariant as well as (mixed) contravariant model tensors (both mixed with different

Kolek, Erik (2024). On the physical foundations of interstellar space travel. In: *Chronicles of Business Informatics Physics (CBIP)*. Volume 2, edition no. 1.0. ISBN: 9783759731159.

numbers of indices) are to be regarded as (two) special cases of the mixed model tensor.

The symmetrical model tensors. The covariant or contravariant model tensor with rank 2 or greater is *not anti-symmetric* if two components, which are created separately by shifting any two specifications (indices), match. A model tensor $A^{\mu\nu}$ or $A_{\mu\nu}$ is therefore *symmetrical* if, with regard to every (conceivable) combination of specifications (indices)

(14) $A^{\mu\nu} = A^{\nu\mu}$ (Einstein, 1916, p. 783),

or

(14a) $A_{\mu\nu} = A_{\nu\mu}$ (Einstein, 1916, p. 783)

corresponds to.

Since the correspondingly conceived model tensor symmetry represents a property {function} that does not depend on the coordinate system, this must be (deduced or) confirmed. With the help of (9), the following is obtained, taking (14) into account

$$A^{\sigma\tau\,'} = (\Omega x_\sigma{}'/\Omega x_\mu) \times (\Omega x_\tau{}'/\Omega x_\nu) \times A^{\mu\nu} = (\Omega x_\sigma{}'/\Omega x_\mu) \times (\Omega x_\tau{}'/\Omega x_\nu) \times A^{\nu\mu} = (\Omega x_\tau{}'/\Omega x_\mu) \times (\Omega x_\sigma{}'/\Omega x_\nu) \times A^{\mu\nu} = A^{\tau\sigma\,'} \text{(Einstein, 1916, p. 783).}$$

After the first and third equal sign, its use is justified by a shift in the summation specifications σ and τ (or μ and ν) (i.e. due to a pure change in the annotation type (comment notation)).

The anti-symmetric model tensors. The covariant or contravariant model tensor with rank 2, 3 or 4 is *anti-symmetric* if two components, which are created separately from each other by shifting any two specifications (indices), *match opposite* each other. A model tensor $A^{\mu\nu}$ or $A_{\mu\nu}$ is therefore *anti-symmetric* if always

(15) $A^{\mu\nu} = -A^{\nu\mu}$ (Einstein, 1916, p. 784),

Kolek, Erik (2024). On the physical foundations of interstellar space travel. In: *Chronicles of Business Informatics Physics (CBIP)*. Volume 2, edition no. 1.0. ISBN: 9783759731159.

or

(15a) $A_{\mu v} = - A_{v\mu}$ (Einstein, 1916, p. 784)

corresponds to.

Of the 16 components $A^{\mu v}$, the twelve components A^{vv} remain (and the four components $A\mu\mu$ are omitted); A^{vv} match as pairs opposite each other, so that (of the 12) only 6 metrically different components exist (six-model vector). In the same way, it is generally to be understood [analogous to (15)] that an anti-symmetric model tensor $A^{\mu v\sigma}$ (with rank 3) has only four metrically different components (four-model vector), (whereas) an anti-symmetric model tensor $A^{\mu v\sigma\tau}$ has only one component. Non anti-symmetric model tensors only exist up to rank 4 within a four-dimensional universe.

§ 7. Multiplication calculation with model tensors.

External multiplication calculation with model tensors. In general, the components of a model tensor with rank z and a second model tensor with rank z' are used to determine the components of the model tensor with rank z + z' by generally multiplying each component of the first tensor with each component of the second tensor per pair. Accordingly, for example, the model tensors T result from the different model tensors A and B

$T_{\mu v\sigma} = A_{\mu v}B_\sigma$ (Einstein, 1916, p. 784),

$T^{\alpha\beta\gamma\delta} = A^{\alpha\beta}B^{\gamma\delta}$ (Einstein, 1916, p. 784),

$T_{\alpha\beta}{}^{\gamma\delta} = A_{\alpha\beta}B^{\gamma\delta}$ (Einstein, 1916, p. 784).

Proof of the coincidence (agreement) of all T with model tensors can be determined directly using the systems of equations (8), (10) and (12) or alternatively using the transformation methods (9), (11) and (13). The representation systems (8), (10) and

Kolek, Erik (2024). On the physical foundations of interstellar space travel. In: *Chronicles of Business Informatics Physics (CBIP)*. Volume 2, edition no. 1.0. ISBN: 9783759731159.

(12) also correspond to examples of external multiplication calculation (with model tensors of rank 1).

{*Internal multiplication calculation with model tensors*. Albert Einstein described equation (6) as the internal multiplication result of a covariant four-model vector A_μ and a contravariant four-model vector A^ν. Equivalently, by means of internal multiplication calculation …}

"Reduction" of mixed model tensors. With the help of mixed model tensors, tensors with ranks of n − 2 can be formed by generally equating a tensor with a covariant specified index and a tensor with a contravariant specified index and summing (or tapering) according to these indices. In general, for example, a mixed model tensor with rank 4 ($A^{\gamma\delta}{}_{\alpha\beta}$) is used to form a mixed model tensor with rank 2

$$A^\delta{}_\beta = A^{\alpha\delta}{}_{\alpha\beta} \left(= \sum_\alpha A^{\alpha\delta}{}_{\alpha\beta} \right) \text{ (Einstein, 1916, p. 785)}$$

and on its basis, again by means of reduction, a model tensor with the rank 0

$$A = A^\beta{}_\beta = A^{\alpha\beta}{}_{\alpha\beta} \text{ (Einstein, 1916, p. 785).}$$

As an evaluation with regard to this result, which was obtained with the help of the reduction, i.e. whether the tensor really represents a tensor, i.e. whether it has a model tensor character, can be checked using the tensor equation corresponding to the generalization in (12), linked with regard to (6) or using the generalization in (13).

Internal and mixed multiplication calculation with model tensors. Both are based on the combination of the external multiplication calculation and the reduction.

For example: – Albert Einstein used the covariant model tensor with rank 2 $A_{\mu\nu}$ and a contravariant model tensor with rank B^σ to form a mixed model tensor by means of an external multiplication calculation

$$D_{\mu\nu}{}^\sigma = A_{\mu\nu}B^\sigma. \text{ (Einstein, 1916, p. 785).}$$

Kolek, Erik (2024). On the physical foundations of interstellar space travel. In: *Chronicles of Business Informatics Physics (CBIP)*. Volume 2, edition no. 1.0. ISBN: 9783759731159.

Reduction according to the specifications σ and ν results in a covariant four-model vector

$$D_\mu = D_{\mu\nu}{}^\nu = A_{\mu\nu}B^\nu \text{ (Einstein, 1916, p. 785)}.$$

Albert Einstein called the resulting four-vector the inner tensor product of $A_{\mu\nu}$ and B^σ. In the same way, the internal tensor product $A_{\mu\nu}B^{\mu\nu}$ is generally formed using the model tensors $A_{\mu\nu}$ and $B^{\sigma\tau}$ by means of an external multiplication calculation and double reduction. The mixed model tensor with rank 2 $D^\tau{}_\mu = A_{\mu\nu}B^{\nu\tau}$ is generally determined using $A_{\mu\nu}$ and $B^{\sigma\tau}$ by means of an external product calculation and a reduction. This calculation procedure is generally referred to as a mixed procedure, as it refers to an external procedure with regard to the values τ and μ and an internal procedure with regard to the values σ and ν.

Albert Einstein now proved an axiom that appears to be mostly useful for the proof of model tensor coincidence. According to the previous description, $A_{\mu\nu}B^{\mu\nu}$ corresponds to an invariant (scalar) if $A_{\mu\nu}$ and $B^{\sigma\tau}$ (also) represent model tensors. However, Albert Einstein also assumed the following. If $A_{\mu\nu}B^{\mu\nu}$ is a scalar with respect to any selection of the model tensor $B^{\mu\nu}$, then $A_{\mu\nu}$ has model tensor coincidence.

Evaluation: – In general, according to the condition regarding an arbitrary change (substitution)

$$A_{\sigma\tau}{}'B^{\sigma\tau'} = A_{\mu\nu}B^{\mu\nu} \text{ (Einstein, 1916, p. 785)}.$$

However, according to the counterpart to (9)

$$B^{\mu\nu} = (\Omega x_\mu/\Omega x_\sigma{}') \times (\Omega x_\nu/\Omega x_\tau{}') \times B^{\sigma\tau'} \text{ (Einstein, 1916, p. 786)}.$$

This, transferred into the previous equation system, results in:

$$[A_{\sigma\tau}{}' - (\Omega x_\mu/\Omega x_\sigma{}') \times (\Omega x_\nu/\Omega x_\tau{}') \times A_{\mu\nu}]B^{\sigma\tau'} = 0 \text{ (Einstein, 1916, p. 786)}.$$

Kolek, Erik (2024). On the physical foundations of interstellar space travel. In: *Chronicles of Business Informatics Physics (CBIP)*. Volume 2, edition no. 1.0. ISBN: 9783759731159.

Due to the arbitrary selection regarding $B^{\sigma\tau}$, this equation is only true (or correct) if the square bracket is resolved, which results in a hypothesis as a sequence, taking into account (11).

This hypothesis is valid as an axiom of agreement with regard to model tensors with arbitrary rank and arbitrary coincidence; an evaluation must always be carried out in the same way.

The axiom can also be proven by the equation: If B^{μ} and C^{ν} correspond to arbitrarily selected vectors (regardless of the vector size, for example, regardless of whether they are four-vectors or six-vectors), then the inner product result corresponds to the following with regard to the arbitrary selection of the same vector

$A_{\mu\nu}B^{\mu}C^{\nu}$ (Einstein, 1916, p. 786)

is an invariant, thus $A_{\mu\nu}$ resembles a covariant model tensor. The previous expression is also still valid in the case where {only seems clear} only a *non-general* statement is true that by arbitrarily choosing a four-model vector B^{μ} the scalar product calculus

$A_{\mu\nu}B^{\mu}B^{\nu}$ (Einstein, 1916, p. 786)

results in an invariant if it is generally known whether $A_{\mu\nu}$ is equal to the symmetry condition $A_{\mu\nu} = A_{\nu\mu}$. Since the method described above generally evaluates the model tensor coincidence with respect to $A_{\mu\nu} + A_{\nu\mu}$, the model tensor agreement with respect to $A_{\mu\nu}$ represents a separate sequence due to the symmetry function. The latter expression can also be easily generalized due to the assumption of covariant and contravariant model tensors independent of their rank.

{An axiom is also valid: If $A_{\mu\nu}B^{\nu}$ represents a model tensor by the arbitrary selection of a four-model vector B^{ν}, then $A_{\mu\nu}$ appears as a model tensor. Due to the requirements or a sufficient evaluation, equivalence to the previous statement is possible.}

Concluding from the evaluation, the expression that cannot be specialized even with respect to arbitrary model tensors becomes conceivable: If {an inner multiplication

Kolek, Erik (2024). On the physical foundations of interstellar space travel. In: *Chronicles of Business Informatics Physics (CBIP)*. Volume 2, edition no. 1.0. ISBN: 9783759731159.

result} forms the values $A_{\mu\nu}B^{\nu}$ by the arbitrary selection of a quad model vector B^{ν} a model tensor with rank 1, then $A_{\mu\nu}$ resembles a model tensor with rank 2. If, for example, C^{μ} resembles an arbitrary four-model vector, then, due to the tensor coincidence $A_{\mu\nu}B^{\nu}$, an inner product operation $A_{\mu\nu}C^{\mu}B^{\nu}$ resembles an invariant through the arbitrary selection of two four-model vectors C^{μ} and B^{ν}, which makes the following theory conceivable.

§ 8. Several properties of the fundamental model tensor $g_{\mu\nu}$.

The fundamental covariant model tensor. Within the scalar theorem on the squaring of a line element system

$$ds^2 = g_{\mu\nu}dx_{\mu}dx_{\nu} \text{ (Einstein, 1916, p. 786)}$$

dx_{μ} has the task of an arbitrarily selectable contravariant model vector. Because $g_{\mu\nu} = g_{\nu\mu}$ also applies, the result of the descriptions in the previous section is that $g_{\mu\nu}$ resembles a covariant model tensor with rank 2. Albert Einstein called this a "fundamental model tensor". Subsequently, Albert Einstein made a derivation of several functions (properties) of the fundamental model tensor, which, however, occur in all tensors with rank 2; however, the special task of the fundamental model tensor in Albert Einstein's theory and Erik Kolek's counter-theory, both of which have the same physical justification in their peculiarity with regard to gravitational interactions. This has the condition that the connections to be conceived only appear as one truth for the fundamental model tensor for Albert Einstein and Erik Kolek (i.e. can only be of importance for an experience to be formed).

The fundamental contravariant model tensor. If a (smaller) sub-determinant (thought) matrix is generally formed within the determinant (thought) matrix of $g_{\mu\nu}$ with respect to each $g_{\nu\mu}$ and the latter is divided by the determination number g equal to $|g_{\mu\nu}|$ of $g_{\mu\nu}$, then generally certain values $g^{\mu\nu}$ (equal to $g^{\nu\mu}$) result, on the basis of which Albert Einstein wanted to evaluate that these resemble a contravariant model tensor.

Kolek, Erik (2024). On the physical foundations of interstellar space travel. In: *Chronicles of Business Informatics Physics (CBIP)*. Volume 2, edition no. 1.0. ISBN: 9783759731159.

According to a generally accepted determinant (thought) expression, the following applies

(16) $g_{\mu\sigma}g^{\nu\sigma} = \delta_\mu{}^\nu$ (Einstein, 1916, p. 787),

here the (mentally interchangeable) symbol $\delta_\mu{}^\nu$ equals either 0 or 1, depending on whether $\mu\nu$ {μ not equal to ν} or μ equals ν. Instead of the previous mathematical thought (theorem) regarding ds^2, Albert Einstein was also allowed to use

$g_{\mu\sigma}\delta_\nu{}^\sigma dx_\mu dx_\nu$ (Einstein, 1916, p. 787),

or according to (16) also

$g_{\mu\sigma}g_{\nu\tau}g^{\sigma\tau}dx_\mu dx_\nu$ (Einstein, 1916, p. 787)

note. Now {exist}, however, according to the multiplication rules from the last section, form the values

$d\xi_\sigma = g_{\mu\sigma}dx_\mu$ (Einstein, 1916, p. 787)

{and $d\xi_\tau = g_{\nu\tau}dx_\nu$} (Einstein, Manuscript, p. 28)

a covariant four-model vector {covariant four-model vectors}, as well as an arbitrarily selectable four-model vector alone (due to the arbitrary selectability with regard to all dx_μ). By adding this to his equation thought (theorem), Albert Einstein determined

$ds^2 = g^{\sigma\tau}d\xi_\sigma d\xi_\tau$ (Einstein, 1916, p. 787).

Because this expression represents an invariant (scalar) due to the arbitrary selection of a model vector $d\xi_\sigma$ and $g^{\sigma\tau}$ appears symmetrical according to the annotated (mental) definition within the specifications (indices) τ and σ, it can be concluded with the help of the results from the last section that $g^{\sigma\tau}$ resembles a contravariant model tensor. With the help of (16), it is (further) conceivable that $\delta_\mu{}^\nu$ also resembles a model tensor which Albert Einstein was (mentally) able to designate as a mixed fundamental model tensor.

Kolek, Erik (2024). On the physical foundations of interstellar space travel. In: *Chronicles of Business Informatics Physics (CBIP)*. Volume 2, edition no. 1.0. ISBN: 9783759731159.

Determination number of a fundamental model tensor. According to the multiplication expression of the determinants, the following applies

$|g_{\mu\alpha}g^{\alpha\nu}| = |g_{\mu\alpha}| \times |g^{\alpha\nu}|$ (Einstein, 1916, p. 788).

On the other hand

$|g_{\mu\alpha}g^{\alpha\nu}| = |\delta_{\mu}{}^{\nu}| = 1$ (Einstein, 1916, p. 788).

Accordingly, the following applies

(17) $|g_{\mu\nu}| \times |g^{\mu\nu}| = 1$ (Einstein, 1916, p. 788).

Volume scales. Albert Einstein first tried to find (i.e. derive) the transformation method with regard to the determinants g equal to $|g_{\mu\nu}|$. According to (11)

$g' = |(\Omega x_{\mu}/\Omega x_{\sigma}{}') \times (\Omega x_{\nu}/\Omega x_{\tau}{}') \times g_{\mu\nu}|$ (Einstein, 1916, p. 788).

This results in the following sequence by using the multiplication expression of the destination numbers twice

$g' = |(\Omega x_{\mu}/\Omega x_{\sigma}{}')| \times |(\Omega x_{\nu}/\Omega x_{\tau}{}')| \times |g_{\mu\nu}| = |(\Omega x_{\mu}/\Omega x_{\sigma}{}')|^{2}g$ (Einstein, 1916, p. 788),

respectively

$\sqrt{g'} = |(\Omega x_{\mu}/\Omega x_{\sigma}{}')|\sqrt{g}$ (Einstein, 1916, p. 788).

On the other hand, the method is similar to the transformation of the volume element system

$d\tau' = \int dx_i dx_i dx_i dx_i$ {$= d\tau$ (Einstein, manuscript, p. 28)} whereby i = 1, 2, 3, 4 for four dimensions (Einstein, 1916, p. 788)

according to the generally accepted expression after Jakobis

$d\tau' = |(\Omega x_{\sigma}{}'/\Omega x_{\mu})|d\tau$ (Einstein, 1916, p. 788).

Kolek, Erik (2024). On the physical foundations of interstellar space travel. In: *Chronicles of Business Informatics Physics (CBIP)*. Volume 2, edition no. 1.0. ISBN: 9783759731159.

By multiplying the two previous systems of equations, the following is generally calculated

(18) $\sqrt{g'} \times d\tau' = \sqrt{g} \times d\tau$ (Einstein, 1916, p. 788) determined.

Instead of $\sqrt{g}$, Albert Einstein subsequently introduced the value $\sqrt{-g}$, which always represents a real (really existing) quantity due to the hyperbolic (non-Euclidean) model tensor coincidence (analogous to that) of our time-space universe. The scalar $\sqrt{-g} \times d\tau$ corresponds to the value of the coordinate system located according to its "position" which is measured on (its) four dimensions analogous to a volume element system {directly} with (practically) rigid (identically constituted) clock hand devices as well as (uniform) test rods according to the (physical) meaning of Albert Einstein's special theory of relativity.

Side note regarding the property of our space-time universe. The condition that the {usual} special theory of relativity always (also) remains the same within the incessantly small makes it possible that ds^2 can always be formulated according to (1a) by means of the real values dX_1, dX_2, dX_3 and dX_4. Albert Einstein described $d\tau_0$ as a {"physically measured"} volume element system dX_1, dX_2, dX_3 and dX_4 that is generally given by "nature". volume element system dX_1 to dX_4, then accordingly

(18a) $d\tau_0 = \sqrt{-g} \times d\tau$ (Einstein, 1916, p. 789).

If $\sqrt{-g}$ is omitted in a region of the universe with four dimensions, then this means that in this space-time a finite reference system volume resembles a non-finite minor volume element system. This could not represent physical reality at any location. Otherwise g would not be able to become negative; according to the special theory of relativity, Albert Einstein assumed that g always has a non-infinite, non-positive value. This represents a theory that simultaneously represents the investigated natural conditions of the physical universe as well as a definition of coordinate point selection.

However, if $-g$ is always non-negative and non-infinite, then it is easy to understand that the point selection must be determined in advance (a posteriori) so that this value

Kolek, Erik (2024). On the physical foundations of interstellar space travel. In: *Chronicles of Business Informatics Physics (CBIP)*. Volume 2, edition no. 1.0. ISBN: 9783759731159.

corresponds to 1. In general, it should be understood later that with the help of such a restriction of the point selection, a meaningful reduction of the complexity of general laws of nature can be achieved. Instead of (18), this results directly in {relationally 18a}

$d\tau' = d\tau$ (Einstein, 1916, p. 789),

from this follows, taking into account the axiom according to Jakobis

(19) $|(\partial x_\sigma'/\partial x_\mu)| = 1$ (Einstein, 1916, p. 789).

According to our point selection, only changes (substitutions) with regard to the point coordinates with the determination number 1 may be carried out.

However, it would be wrong to assume that this approach entails a partial invalidity of the general principle of relativity. Albert Einstein did not want to know: "How he should designate the laws of nature that are covariant with respect to every transformation of determinant 1?" Instead, Albert Einstein wanted to know: "How he should designate the non-specifically (existing) covariant laws of nature?" Albert Einstein first derived these laws of nature, and then reduced the complexity of this formulation by means of a special choice of coordinate system.

Design of previously non-existent model tensors using a fundamental model tensor. By means of the internal, external and mixed multiplication calculation of a model tensor with an underlying (fundamental) model tensor, model tensors with different properties and ranks are formed.

For example:

$A^\mu = g^{\mu\sigma}A_\sigma$ (Einstein, 1916, p. 790),

$A = g_{\mu\nu}A^{\mu\nu}$ (Einstein, 1916, p. 790).

In particular, Albert Einstein wanted to draw attention to the next designs:

Kolek, Erik (2024). On the physical foundations of interstellar space travel. In: *Chronicles of Business Informatics Physics (CBIP)*. Volume 2, edition no. 1.0. ISBN: 9783759731159.

$A^{\mu\nu} = g^{\mu\alpha}g^{\nu\beta}A_{\alpha\beta}$ (Einstein, 1916, p. 790),

$A_{\mu\nu} = g_{\mu\alpha}g_{\nu\beta}A^{\alpha\beta}$ (Einstein, 1916, p. 790)

("completion" of a contravariant relationwise covariant model tensor) and

$B_{\mu\nu} = g_{\mu\nu}g^{\alpha\beta}A_{\alpha\beta}$ (Einstein, 1916, p. 790).

Albert Einstein described $B_{\mu\nu}$ as a complexity-reduced model tensor with respect to $A_{\mu\nu}$. In the same way

$B^{\mu\nu} = g^{\mu\nu}g_{\alpha\beta}A^{\alpha\beta}$ (Einstein, 1916, p. 790).

It should be understood that $g^{\mu\nu}$ resembles a completion of $g_{\mu\nu}$. Since in general

$g^{\mu\alpha}g^{\nu\beta}g_{\alpha\beta} = g^{\mu\alpha}\delta_\alpha{}^\nu = g^{\mu\nu}$ (Einstein, 1916, p. 790).

The following therefore applies to complete the relationship

$g_{\mu\alpha}g_{\nu\beta}g^{\alpha\beta} = g_{\mu\alpha}\delta^\alpha{}_\nu = g_{\mu\nu}$ (Einstein, 1916, p. 790).

§ 9. System of equations for the geodesic straight line (relations of a coordinate movement).

Because a "reference element" *ds* represents an {entirely natural} imaginary space that does not depend on a reference system, a {general} straight line {intended} between the two point neighbors P_1 and P_2 of the universe with four dimensions, with respect to which $\int ds$ represents an extreme value (geodesic straight line), also has a meaning that does not depend on a coordinate selection. The system of equations for both points is

(20) $\delta\{_{P_1}\!\int^{P_2}(ds)\} = 0$ (Einstein, 1916, p. 790).

With the help of this system of equations, four absolute differential equation systems are generally found in a familiar way by carrying out the variation, which determine our geodesic straight line; {also} the deduction described is listed here as a

Kolek, Erik (2024). On the physical foundations of interstellar space travel. In: *Chronicles of Business Informatics Physics (CBIP)*. Volume 2, edition no. 1.0. ISBN: 9783759731159.

supplement. λ represents a property of the point coordinates x_v; it forms a number of surfaces that intersect the geodesic line to be found and every line that is not finitely adjacent to it, between the points P_1 and P_2. All curves determined in this way can therefore be thought of as conditional, so that their point coordinates x_v can be formulated within the property of λ. A symbol δ resembles the point of intersection with respect to the first point of the geodesic line to be found with respect to the point on an adjacent trajectory curve that is assigned with respect to the exemplary λ. (20) can then be replaced by a system of equations with regard to the point properties λ_1 and λ_2.

$$\text{(20a)} \quad {}_{\lambda_1}\!\int^{\lambda_2} \delta w \, d\lambda = 0 \text{ (Einstein, 1916, p. 791)}$$
$$w^2 = g_{\mu\nu} \times (dx_\mu/d\lambda) \times (dx_\nu/d\lambda) \text{ (Einstein, 1916, p. 791).}$$

However, because

$$\delta w = 1/w\{1/2 \times (\Omega g_{\mu\nu}/\Omega x_\sigma) \times (dx_\mu/d\lambda) \times (dx_\nu/d\lambda) \times \delta x_\sigma + g_{\mu\nu} \times (dx_\mu/d\lambda) \times \delta(dx_\nu/d\lambda)\}$$
(Einstein, 1916, p. 791),

then generally results according to the incorporation of δw into (20a), taking into account the sequence, so that

$$\delta(dx_\nu/d\lambda) = d\delta x_\nu/d\lambda \text{ (Einstein, 1916, p. 791),}$$

corresponding partial merging (integration) with regard to the point properties λ_1 and λ_2

$$\text{(20b)} \quad {}_{\lambda_1}\!\int^{\lambda_2} d\lambda \varkappa_\sigma \delta x_\sigma = 0 \text{ (Einstein, 1916, p. 791)}$$
$$\varkappa_\sigma = d/d\lambda\{(g_{\mu\nu}/w) \times (dx_\mu/\Omega\lambda)\} - 1/2w \times (\Omega g_{\mu\nu}/\Omega x_\sigma) \times (dx_\mu/d\lambda) \times (dx_\nu/d\lambda)$$
$$\text{(Einstein, 1916, p. 791)}$$

This leads to the elimination of $\varkappa_\sigma$ due to the arbitrary selectability of δx_σ. Accordingly

$$\text{(20c)} \quad \varkappa_\sigma = 0 \text{ (Einstein, 1916, p. 791)}$$

Kolek, Erik (2024). On the physical foundations of interstellar space travel. In: *Chronicles of Business Informatics Physics (CBIP)*. Volume 2, edition no. 1.0. ISBN: 9783759731159.

the equation systems of a geodesic straight line. If *ds is not equal to 0* on the geodesic line under investigation, then Albert Einstein was able to select the "unit arc length" s measured on this geodesic line as the property (parameter, function) λ. This results in *w equal to 1*, and in general, instead of (20c)

$g_{\mu\nu} \times (d^2x_\mu/ds^2) + (\Omega g_{\mu\nu}/\Omega x_\sigma) \times (dx_\sigma/d\lambda) \times (dx_\mu/d\lambda) - 1/2 \times (\Omega g_{\mu\nu}/\Omega x_\sigma) \times (dx_\mu/d\lambda) \times (dx_\nu/d\lambda) = 0$ (Einstein, 1916, p. 791),

or by simply reformulating the system of equations (formulation method)

(20d) $g_{\alpha\sigma} \times (d^2x_\alpha/ds^2) + [\mu\,\nu\mid\sigma] \times (dx_\mu/ds) \times (dx_\nu/ds) = 0$ (Einstein, 1916, p. 791),

or defined according to Christoffel, the equation

(21) $[\mu\,\nu\mid\sigma] = 1/2(\Omega g_{\mu\sigma}/\Omega x_\nu + \Omega g_{\nu\sigma}/\Omega x_\mu - \Omega g_{\mu\nu}/\Omega x_\sigma)$ (Einstein, 1916, p. 791).

If (20d) is now generally multiplied by $g^{\sigma\tau}$ at the end (outer calculation with respect to τ, inner calculation with respect to σ), then the following generally results as the final expression of the system of equations of the geodesic straight line

(22) $d^2x_\tau/ds^2 + \{\mu\,\nu\mid\tau\} \times (dx_\mu/ds) \times (dx_\nu/ds) = 0$ (Einstein, 1916, p. 792).

According to Christoffel, the equation appears to be fixed

(23) $\{\mu\,\nu\mid\tau\} = g^{\tau\alpha}[\mu\,\nu\mid\alpha]$ (Einstein, 1916, p. 792).

§ 10. The design of model tensors using differentiation.

With the help of the system of equations relating to the geodesic straight line, Albert Einstein was now able to easily derive the (necessary) axioms according to which other model tensors can be formed by means of differentiation using model tensors. This enabled Albert Einstein to derive general covariant differential equation systems. Albert Einstein achieved this objective by repeatedly using {the} next comprehensible axiom {approach}.

Kolek, Erik (2024). On the physical foundations of interstellar space travel. In: *Chronicles of Business Informatics Physics (CBIP)*. Volume 2, edition no. 1.0. ISBN: 9783759731159.

If the orbital curve exists in this universe, the coordinates of which are described by means of the unit arc distance s starting from a fixed point on this curve, φ also corresponds to a scalar spatial property, then $d\varphi/ds$ also corresponds to a scalar. The evaluation is carried out in such a way that $d\varphi$ and ds correspond to scalars.

Because

$d\varphi/ds = \Omega\varphi/\Omega x_\mu \times dx_\mu/ds$ (Einstein, 1916, p. 792),

then also equals

$\psi = \Omega\varphi/\Omega x_\mu \times dx_\mu/ds$ (Einstein, 1916, p. 792)

a scalar, but also with regard to every curve that starts from a single starting point in the universe, i.e. with regard to an arbitrary selection of a model vector for the dx_μ. This directly results in the consequence that

(24) $A_\mu = \Omega\varphi/\Omega x_\mu$ (Einstein, 1916, p. 792)

represents a covariant four-model vector (the gradient field with respect to φ).

According to this axiom, the differential division result in the curve also equals

$X = d\psi/ds$ (Einstein, 1916, p. 792)

a scalar. By integrating (inserting) the ψ, Albert Einstein first obtained

$X = \Omega^2\varphi/\Omega x_\mu \Omega x_\nu \times dx_\mu/ds \times dx_\nu/ds + \Omega\varphi/\Omega x_\mu \times d^2x_\mu/ds^2$ (Einstein, 1916, p. 792).

For the time being (i.e. the confirmation), this does not prove the existence of the model tensor coincidence. However, if Albert Einstein now considered that this trajectory, on which he differentiated, corresponds to a geodesic straight line, then he obtained according to (22) by inserting the expression d^2x_ν / ds^2:

$X = \{\Omega^2\varphi/\Omega x_\mu \Omega x_\nu - \{\mu\,\nu\,|\,\tau\} \times \Omega\varphi/\Omega x_\tau\} \times dx_\mu/ds \times dx_\nu/ds$ (Einstein, 1916, p. 793).

Kolek, Erik (2024). On the physical foundations of interstellar space travel. In: *Chronicles of Business Informatics Physics (CBIP)*. Volume 2, edition no. 1.0. ISBN: 9783759731159.

Due to the substitutability of the differentiation equations according to ν and μ as well as from this, so that according to (21) and (23) the curved bracket notation $\{\mu\,\nu\,|\,\tau\}$ is *not anti-symmetric* with respect to μ and ν, it follows that the set of brackets in μ and ν is *not anti-symmetric*. Since it is generally possible to draw a geodesic straight line starting from any starting point in the universe in arbitrary directional notation and $dx_\mu\,/\,ds$ therefore represents a four-model vector with an unrestrictedly selectable size ratio of the components, we obtain as a consequence according to the results in section 7 that

$$(25)\quad A_{\mu\nu} = \Omega^2\varphi/\Omega x_\mu \Omega x_\nu - \{\mu\,\nu\,|\,\tau\} \times \Omega\varphi/\Omega x_\tau \text{ (Einstein, 1916, p. 793)}$$

represents a covariant model tensor with rank 2. Albert Einstein had therefore achieved the result: Using a covariant model tensor with rank 1 {four-model vector}

$$A_\mu = \Omega\varphi/\Omega x_\mu \text{ (Einstein, 1916, p. 793)}$$

he designed a covariant model tensor with rank 2 by means of differentiation

$$(26)\quad A_{\mu\nu} = \Omega A_\mu/\Omega x_\nu - \{\mu\,\nu\,|\,\tau\} \times A_\tau \text{ (Einstein, 1916, p. 793)}.$$

Albert Einstein described the model tensor $A_{\mu\nu}$ as the "completion" of the model tensor A_μ. At first, it was easy for him to determine that its shape also leads to a model tensor if the model vector A_μ does not resemble a (representable) gradient. To understand this, he first learned that

$$\psi(\Omega\varphi/\Omega x_\mu) \text{ (Einstein, 1916, p. 793)}$$

resembles a covariant four-model vector if ψ and φ are invariants. This also applies to a summation described with four corresponding expressions (members)

$$S_\mu = \psi^{(1)}(\Omega\varphi^{(1)}/\Omega x_\mu) + \psi^{(2)}(\Omega\varphi^{(2)}/\Omega x_\mu) + \psi^{(3)}(\Omega\varphi^{(3)}/\Omega x_\mu) + \psi^{(4)}(\Omega\varphi^{(4)}/\Omega x_\mu) \text{ (Einstein,}$$
1916, p. 793),

if $\psi^{(1)}\varphi^{(1)}$, $\psi^{(2)}\varphi^{(2)}$, $\psi^{(3)}\varphi^{(3)}$ and $\psi^{(4)}\varphi^{(4)}$ represent invariants. However, it must now be understood that all covariant four-model vectors can be mapped with the shape S_μ. If,

Kolek, Erik (2024). On the physical foundations of interstellar space travel. In: *Chronicles of Business Informatics Physics (CBIP)*. Volume 2, edition no. 1.0. ISBN: 9783759731159.

for example, A_μ represents a four-model vector and its components represent arbitrarily existing properties of x_v, then in general (with regard to the selected reference system) only the following must be set

$\psi^{(1)} = A_1$ (Einstein, 1916, p. 794), $\qquad \varphi^{(1)} = x_1$ (Einstein, 1916, p. 794),

$\psi^{(2)} = A_2$ (Einstein, 1916, p. 794), $\qquad \varphi^{(2)} = x_2$ (Einstein, 1916, p. 794),

$\psi^{(3)} = A_3$ (Einstein, 1916, p. 794), $\qquad \varphi^{(3)} = x_3$ (Einstein, 1916, p. 794),

$\psi^{(4)} = A_4$ (Einstein, 1916, p. 794), $\qquad \varphi^{(4)} = x_4$ (Einstein, 1916, p. 794),

is achieved so that A_μ matches S_μ.

In order to evaluate, based on this objective, that $A_{\mu v}$ equals a model tensor if an arbitrarily covariant four-model vector should be inserted on the right with respect to A_μ, Albert Einstein only had to prove that this is true with respect to the four-model vector S_μ. With regard to this purpose, however, {since the right-hand side} appears {straight and uniform with respect to A_τ and $\Omega A_{\mu v}/\Omega x_v$} from (26), as a consideration of the right-hand side of the equation shows, it seems satisfactory to evaluate with respect to the scenario

$A_\mu = \psi(\Omega\varphi/\Omega x_\mu)$ (Einstein, 1916, p. 794)

must be made clear. The right-hand term of (25) now has model tensor coincidence, since it has been multiplied by ψ.

$\psi(\Omega^2\varphi/\Omega x_\mu\Omega x_v) - \{\mu\ v\mid\tau\}(\psi\times\Omega\varphi/\Omega x_\tau)$ (Einstein, 1916, p. 794)

Also equals

$\Omega\psi/\Omega x_\mu \times \Omega\varphi/\Omega x_v$ (Einstein, 1916, p. 794)

a model tensor (external product result of two four-model vectors). By means of addition calculation, the model tensor coincidence results as a consequence with regard to

$\Omega/\Omega x_v(\psi\times\Omega\varphi/\Omega x_\mu) - \{\mu\ v\mid\tau\}(\psi\times\Omega\varphi/\Omega x_\tau)$ (Einstein, 1916, p. 794).

Kolek, Erik (2024). On the physical foundations of interstellar space travel. In: *Chronicles of Business Informatics Physics (CBIP)*. Volume 2, edition no. 1.0. ISBN: 9783759731159.

Hereby, as a consideration of (26) makes clear, the required evaluation with regard to the four-model vector

$\psi \times \Omega\varphi/\Omega x_\mu$ (Einstein, 1916, p. 794),

and therefore according to the previously evaluated with regard to each arbitrary four-model vector A_μ. –

Based on the completion of a four-model vector, it is generally easy to think of the "completion" of a (generally) covariant model tensor with arbitrary rank; the described design is like a generalization of the completion of the four-model vector. Albert Einstein concentrated on the derivation {investigation} of the completion of the model tensor with rank 2, because this choice already makes the design law (design science law) easy to understand.

It was learned that all covariant model tensors with rank 2 can be mapped as the summation of model tensors of the type $A_\mu B_\nu$.

The following model tensor and its components are formed using the external multiplication calculation of four-vectors in conjunction with (arbitrarily existing) components A_{11} to A_{14} or 1 and 3×0

A_{11}	A_{12}	A_{13}	A_{14}	
0	0	0	0	
0	0	0	0	
0	0	0	0	(Einstein, 1916, p. 795).

With the help of the addition calculation of four such model tensors, the model tensor $A_{\mu\nu}$ and its arbitrarily defined components are generally (mentally) summarized.

Therefore, it must be sufficient to set up the formulation of the completion with respect to a corresponding specialized model tensor. According to (26), the following formulations have

$\Omega A_\mu / \Omega x_\sigma - \{\sigma\,\mu \mid \tau\} A_\tau$ (Einstein, 1916, p. 795),

$\Omega B_v / \Omega x_\sigma - \{\sigma\,v \mid \tau\} B_\tau$ (Einstein, 1916, p. 795)

model tensor coincidence. An external multiplication calculation of the first-place expression with B_v and the second-place expression with A_μ generally results in a model tensor with rank 3; the addition calculation from both model tensors with rank 3 leads to the (lower) model tensor with rank 3

(27) $A_{\mu v \sigma} = \Omega A_{\mu v} / \Omega x_\sigma - \{\sigma\,\mu \mid \tau\} A_{\tau v} - \{\sigma\,v \mid \tau\} A_{\mu\tau}$ (Einstein, 1916, p. 795),

where $A_{\mu v}$ is fixed equal to $A_\mu B_v$. Because the right-hand term in (27) is straightforward and uniform with respect to $A_{\mu v}$ and its initial derivatives. The described design law directs, both in the case of a model tensor of the type $A_\mu B_v$ and also in the case of the summation of these model tensors, i.e. in the case of an arbitrarily covariant model tensor with rank 2, towards the model tensor. Albert Einstein described $A_{\mu v \sigma}$ as the completion of the model tensor $A_{\mu v}$.

It seems plausible that (24) and (26) only represent *non-general* equation scenarios of (27) (completion of the model tensor with the rank 1 or 0). Overall, every *non-general* design law of model tensors can be understood on the basis of (27) in conjunction with model tensor multiplication calculation.

§ 11. Several non-general scenarios with high usefulness.

Several simplification theorems {differential theorems} with regard to the fundamental model tensor. Albert Einstein first devised or derived several simplification theorems (support theorems), which were later often used as systems of equations. In accordance with the law of differentiation of determinants, the following applies

(28) $dg = g^{\mu v} g \, dg_{\mu v} = - g_{\mu v} g \, dg^{\mu v}$ (Einstein, 1916, p. 796).

Kolek, Erik (2024). On the physical foundations of interstellar space travel. In: *Chronicles of Business Informatics Physics (CBIP)*. Volume 2, edition no. 1.0. ISBN: 9783759731159.

This equation is justified by the previous equation if it is generally remembered that $g_{\mu\nu}g^{\mu'\nu} = \delta_\mu{}^{\mu'}$ and therefore $g_{\mu\nu}g^{\mu\nu} = 4$, from which follows

$g_{\mu\nu}dg^{\mu\nu} + g^{\mu\nu}dg_{\mu\nu} = 0$ (Einstein, 1916, p. 796).

With the help of (28)

(29) $1/\sqrt{-g} \times \Omega\sqrt{-g}/\Omega x_\sigma = 1/2 \times \Omega lg(-g)/\Omega x_\sigma = 1/2 \times g^{\mu\nu} \times \Omega g_{\mu\nu}/\Omega x_\sigma = -1/2 \times g_{\mu\nu} \times \Omega g^{\mu\nu}/\Omega x_\sigma$ (Einstein, 1916, p. 796).

Starting from

$g_{\mu\sigma}g^{\nu\sigma} = \delta_\mu{}^\nu$ (Einstein, 1916, p. 796)

additionally produces the sequence

$$(30) \quad \left\{ \begin{array}{l} \text{relations} \\ \text{wise} \end{array} \right. \quad \begin{array}{l} g_{\mu\sigma}dg^{\nu\sigma} = -g^{\nu\sigma}dg_{\mu\sigma} \\[4pt] g_{\mu\sigma} \times \Omega g^{\nu\sigma}/\Omega x_\lambda = -g^{\nu\sigma} \times \Omega g_{\mu\sigma}/\Omega x_\lambda \end{array} \quad \text{(Einstein, 1916, p. 796).}$$

By means of a mixed multiplication calculation using $g^{\sigma\tau}$ or $g_{\nu\lambda}$, this generally becomes (in the case of a modified formulation of the information)

$$(31) \quad \left\{ \begin{array}{l} dg^{\mu\nu} = -g^{\mu\alpha}g^{\nu\beta}dg_{\alpha\beta} \\[4pt] \Omega g^{\mu\nu}/\Omega x_\sigma = -g^{\mu\alpha}g^{\nu\beta} \times \Omega g_{\alpha\beta}/\Omega x_\sigma \end{array} \right. \quad \text{(Einstein, 1916, p. 796),}$$

relations wise

$$(32) \quad \left\{ \begin{array}{l} dg_{\mu\nu} = -g_{\mu\alpha}g_{\nu\beta}dg^{\alpha\beta} \\[4pt] \Omega g_{\mu\nu}/\Omega x_\sigma = -g_{\mu\alpha}g_{\nu\beta} \times \Omega g^{\alpha\beta}/\Omega x_\sigma \end{array} \right. \quad \text{(Einstein, 1916, p. 796).}$$

A reference (31) enables the transformation, which Albert Einstein also had to use several times. According to (21)

(33) $\Omega g_{\alpha\beta}/\Omega x_\sigma = [\alpha\,\sigma\,|\,\beta] + [\beta\,\sigma\,|\,\alpha]$ (Einstein, 1916, p. 796).

Kolek, Erik (2024). On the physical foundations of interstellar space travel. In: *Chronicles of Business Informatics Physics (CBIP)*. Volume 2, edition no. 1.0. ISBN: 9783759731159.

If this expression is generally inserted into the lower of the equations (31), then the following generally follows, taking into account (23)

(34) $\Omega g_{\mu\nu}/\Omega x_\sigma = - (g^{\mu\tau}\{\tau\,\sigma\,|\,\nu\} + g^{\nu\tau}\{\tau\,\sigma\,|\,\mu\})$ (Einstein, 1916, p. 796).

The change in the right-hand term in (34) with an effect on (29) generally results in

(29a) $1/\sqrt{-y} \times \Omega\sqrt{-g}/\Omega x_\sigma = \{\mu\,\sigma\,|\,\mu\}$ (Einstein, 1916, p. 796).

The divergence of a contravariant four-model vector. If (26) is generally multiplied by a contravariant fundamental model tensor $g^{\mu\nu}$ (internal multiplication calculation), then the right-hand term after its transformation of its first-place expression first receives the notation

$\Omega/\Omega x_\nu(g^{\mu\nu}A_\mu) - A_\mu \times \Omega g^{\mu\nu}/\Omega x_\nu - 1/2 \times g^{\tau\alpha}(\Omega g^{\mu\alpha}/\Omega x_\nu + \Omega g_{\nu\alpha}/\Omega x_\mu - \Omega g_{\mu\nu}/\Omega x_\alpha)g^{\mu\nu}A_\tau$
(Einstein, 1916, p. 797).

According to (29) and (31), the latter part of this equation allows the expression

$1/2 \times \Omega g^{\tau\nu}/\Omega x_\nu \times A_\tau + 1/2 \times \Omega g^{\tau\mu}/\Omega x_\mu \times A_\tau + 1/\sqrt{-g} \times \Omega\sqrt{-g}/\Omega x_\alpha \times g^{\mu\nu}A_\tau$ (Einstein, 1916, p. 797).

Because the designation of the sums is unimportant, the first two expressions of this equation differ from the second expression of the above equation; the latter expression can be merged with the first expression of the above equation.

If it is generally determined that

$g^{\mu\nu}A_\mu = A^\nu$ (Einstein, 1916, p. 797)

is, where A_μ is an arbitrarily selectable model vector just like A^ν, then the following is generally determined at the end

(35) $\Phi = 1/\sqrt{-g} \times \Omega/\Omega x_\nu (\sqrt{-g} \times A^\nu)$ (Einstein, 1916, p. 797).

This invariant is similar to the repulsion (divergence) with respect to the contravariant four-model vector {model tensor} A^v.

"Rotation" of a (covariant) four-model vector. The second expression in (26) does not correspond to any *anti-symmetry* in its specifications (indices) μ and v. Therefore, $A_{\mu v} - A_{v\mu}$ resembles a very simple (*non-symmetric*) model tensor. Generally follows

(36) $B_{\mu v} = \Omega A_\mu / \Omega x_v - \Omega A_v / \Omega x_\mu$ (Einstein, 1916, p. 797).

Non-symmetrical (non-uniform) completion of the six-model vector. If (27) is generally applied to a model tensor with missing symmetry (*anti-symmetry*) and of rank 2 $A_{\mu v}$, the two systems of equations derived by means of periodic (cyclic) displacement of the specifications μ, v and σ as well as the three systems of equations are added, then the model tensor with rank 3 generally follows

(37) $B_{\mu v \sigma} = A_{\mu v \sigma} + A_{v \sigma \mu} + A_{\sigma \mu v} = \Omega A_{\mu v}/\Omega x_\sigma + \Omega A_{v\sigma}/\Omega x_\mu + \Omega A_{\sigma\mu}/\Omega x_v$ (Einstein, 1916, p. 797),

with which it can be easily evaluated that this symmetry is missing (i.e. an *anti-symmetry* can be proven).

Deviation (divergence) of the six-model vector. If (27) is generally multiplied by $g^{\mu\alpha}g^{v\beta}$ (mixed multiplication calculation), then a model tensor is generally also formed. The first expression of the right-hand term in (27) may generally be expressed as

$\Omega/\Omega x_\sigma (g^{\mu\alpha}g^{v\beta}A_{\mu v}) - g^{\mu\alpha} \times \Omega g^{v\beta}/\Omega x_\sigma \times A_{\mu v} - g^{v\beta} \times \Omega g^{\mu\alpha}/\Omega x_\sigma \times A_{\mu v}$ (Einstein, 1916, p. 798)

can be described. If the substitution of $g^{\mu\alpha}g^{v\beta}A_{\mu v\sigma}$ with $A_\sigma{}^{\alpha\beta}$, $g^{\mu\alpha}g^{v\beta}A_{\mu v}$ with $A^{\alpha\beta}$ is carried out in general and the following is generally changed within the transformed former expression

$\Omega g^{v\beta}/\Omega x_\sigma$ AND $\Omega g^{\mu\alpha}/\Omega x_\sigma$ (Einstein, 1916, p. 798)

Kolek, Erik (2024). On the physical foundations of interstellar space travel. In: *Chronicles of Business Informatics Physics (CBIP)*. Volume 2, edition no. 1.0. ISBN: 9783759731159.

with respect to (34), then the right-hand term in (27) is transformed into a system of equations with seven members, within which four components differ. This results in a remainder

(38) $A_\sigma{}^{\alpha\beta} = \Omega A^{\alpha\beta}/\Omega x_\sigma + \{\sigma\varkappa \mid \alpha\} A^{\varkappa\beta} + \{\sigma\varkappa \mid \beta\} A^{\alpha\varkappa}$ (Einstein, 1916, p. 798).

This corresponds to the system of equations with regard to the completion of a contravariant model tensor with rank 2, which can also be derived analogously with regard to contravariant model tensors with lower and higher rank.

Albert Einstein now understood that the same method could also be used to design the completion of a mixed model tensor $A_\mu{}^\alpha$:

(39) $A_{\mu\sigma}{}^\alpha = \Omega A^\alpha{}_\mu/\Omega x_\sigma - \{\sigma\mu \mid \tau\} A^\alpha{}_\tau + \{\sigma\tau \mid \alpha\} A^\tau{}_\mu$ (Einstein, 1916, p. 798).

By reducing (38) with regard to the specifications β and σ (internal multiplication calculation using $\delta^\sigma{}_\beta$), the contravariant four-model vector generally follows

$A^\alpha = \Omega A^{\alpha\beta}/\Omega x_\beta + \{\beta\varkappa \mid \beta\} A^{\alpha\varkappa} + \{\beta\varkappa \mid \alpha\} A^{\varkappa\beta}$ (Einstein, 1916, p. 798).

Due to the non-missing symmetry of the content of the curly bracket $\{\beta\varkappa \mid \alpha\}$ with respect to the specifications β and $\varkappa$, the third part of the right-hand term is resolved if $A^{\alpha\beta}$ represents a model tensor with missing symmetry (*anti-symmetry*), which Albert Einstein wanted to assume; the second part can be rearranged accordingly according to (29a). Albert Einstein thus obtained

(40) $A^\alpha = 1/\sqrt{-g} \times \Omega(\sqrt{-g} \times A^{\alpha\beta})/\Omega x_\beta$ (Einstein, 1916, p. 798).

This corresponds to the equation for the divergence with respect to a contravariant {*anti-symmetric*} six-model vector {model tensor with rank 2}.

The deviation (divergence) of a mixed model tensor with rank 2. If Albert Einstein designed the reduction of (39) with regard to the specifications α and σ, then, taking (29a) into account, he obtained

Kolek, Erik (2024). On the physical foundations of interstellar space travel. In: *Chronicles of Business Informatics Physics (CBIP)*. Volume 2, edition no. 1.0. ISBN: 9783759731159.

(41) $\sqrt{-g} \times A_\mu = \Omega(\sqrt{-g} \times A_\mu{}^\sigma)/\Omega z_\sigma - \{\sigma \, \mu \mid \tau\}\sqrt{-g} \times A_\tau{}^\sigma$ (Einstein, 1916, p. 799).

If the contravariant model tensor $A^{\varrho\sigma} = g^{\varrho\tau}A_\tau{}^\sigma$ is generally inserted in the latter component, then this is given the expression

$- \{\sigma \, \mu \mid \varrho\}\sqrt{-g} \times A^{\varrho\sigma}$ (Einstein, 1916, p. 799).

If the model tensor $A^{\varrho\sigma}$ additionally does not resemble an *anti-symmetric* one, then its shape reduces to

$- 1/2\sqrt{-g} \times \Omega g_{\varrho\sigma}/\Omega x_\mu \times A^{\varrho\sigma}$ (Einstein, 1916, p. 799).

If instead of $A^{\varrho\sigma}$ the also not *anti-symmetric* covariant model tensor $A_{\varrho\sigma} = g_{\varrho\alpha}g_{\sigma\beta}A^{\alpha\beta}$ were added, then the latter part would have the expression

$1/2\sqrt{-g} \times \Omega g^{\varrho\sigma}/\Omega x_\mu \times A_{\varrho\sigma}$ (Einstein, 1916, p. 799)

is assumed. Within this (mentally) observed symmetry truth, (41) may therefore also be assumed by means of the two expressions

(41a) $\sqrt{-g} \times A_\mu = \Omega(\sqrt{-g} \times A_\mu{}^\sigma)/\Omega x_\sigma - 1/2 \times \Omega g_{\varrho\sigma}/\Omega x_\mu \times \sqrt{-g} \times A^{\varrho\sigma}$ (Einstein, 1916, p. 799)

and

(41b) $\sqrt{-g} \times A_\mu = \Omega(\sqrt{-g} \times A_\mu{}^\sigma)/\Omega x_\sigma + 1/2 \times \Omega g^{\varrho\sigma}/\Omega x_\mu \times \sqrt{-g} \times A_{\sigma\varrho}$ (Einstein, 1916, p. 799)

exchange took place, which Albert Einstein later had to use.

§ 12. The model tensor according to Riemann-Christoffel.

Albert Einstein was now interested in those model tensors that can only be determined using the fundamental model tensor of all $g_{\mu\nu}$ by means of differentiation calculation. A derivation for this appears to be quite simple at first. In (27), instead of an arbitrarily selected model tensor $A_{\mu\nu}$, he inserted a fundamental model tensor of all $g_{\mu\nu}$ in such a

Kolek, Erik (2024). On the physical foundations of interstellar space travel. In: *Chronicles of Business Informatics Physics (CBIP)*. Volume 2, edition no. 1.0. ISBN: 9783759731159.

way that he obtained a modified (renewed) model tensor, the completion of the fundamental model tensor, so to speak. However, he had simply learned that this completion can be resolved analogously. However, he achieved his goal with the next method. For this he added in (27)

$A_{\mu\nu} = \Omega A_\mu / \Omega x_\nu - \{\mu\ \nu \mid \varrho\} A_\varrho$ (Einstein, 1916, p. 799),

added a completion of the four-model vector A_ν. It then received the model tensor with rank 3 (with a minimally updated designation of the data)

$$A_{\mu\sigma\tau} = \Omega^2 A_\mu / \Omega x_\sigma \Omega x_\tau$$
$$- \{\mu\ \sigma \mid \varrho\}\Omega A_\varrho / \Omega x_\tau - \{\mu\ \tau \mid \varrho\}\Omega A_\varrho / \Omega x_\sigma - \{\sigma\ \tau \mid \varrho\}\Omega A_\mu / \Omega x_\varrho$$
$$+ [- \Omega / \Omega x_\tau \{\mu\ \sigma \mid \varrho\} + \{\mu\ \tau \mid \alpha\}\{\alpha\ \sigma \mid \varrho\} + \{\sigma\ \tau \mid \alpha\}\{\alpha\ \mu \mid \varrho\}]A_\varrho \text{ (Einstein,}$$
$$\text{1916, p. 800)}.$$

This system of equations simplifies the design of the model tensor $A_{\mu\sigma\tau} - A_{\mu\tau\sigma}$. Since the next expressions of the equation with regard to $A_{\mu\sigma\tau}$ can be resolved with those for $A_{\mu\tau\sigma}$: the expression number 1, the expression number 4, and the expression which corresponds to the notation at the end within the bracket with corners; since each of these expressions within the σ and τ has no *anti-symmetry*. This applies analogously to the summation of the expressions in the second and third position. Albert Einstein therefore obtained

(42) $A_{\mu\sigma\tau} - A_{\mu\tau\sigma} = B_{\mu\sigma\tau}{}^\varrho A_\varrho$ (Einstein, 1916, p. 800),

$$B_{\mu\sigma\tau}{}^\varrho = - \Omega / \Omega x_\tau \{\mu\ \sigma \mid \varrho\} + \Omega / \Omega x_\sigma \{\mu\ \tau \mid \varrho\}$$

(43) $\{$
$$- \{\mu\ \sigma \mid \alpha\}\{\alpha\ \tau \mid \varrho\} + \{\mu\ \tau \mid \alpha\}\{\alpha\ \sigma \mid \varrho\} \text{ (Einstein, 1916, p.}$$
$$\text{800)}.$$

Based on this result, it is important to recognize that only all A_ϱ but none of their deductions (constellations, derivatives) can be found on the right-hand side of (42). Because of the model tensor coincidence (tensor correspondence) of $A_{\mu\sigma\tau} - A_{\mu\tau\sigma}$ combined with the fact that A_ϱ represents an arbitrarily selectable four-model vector,

Kolek, Erik (2024). On the physical foundations of interstellar space travel. In: *Chronicles of Business Informatics Physics (CBIP)*. Volume 2, edition no. 1.0. ISBN: 9783759731159.

it appears as a consequence, conceivable from the results in section 7, that $B_{\mu\sigma\tau}{}^{\varrho}$ represents a model tensor (the model tensor according to Riemann-Christoffel).

The sense (purely) according to the mathematics of the Riemann-Christoffel tensor is justified as follows: If our universe should be designed in such a way that a line element system (reference system) exists, with respect to which all $g_{\mu v}$ represent invariables, then every $R_{\mu\sigma\tau}{}^{\varrho}$ dissolves. If an arbitrarily selected system is generally rethought instead of the initial volume element system, then the $g_{\mu v}$ assigned to the new system cannot represent variables (or constant variables). However, the model tensor correspondence (tensor coincidence) of the $R_{\mu\sigma\tau}{}^{\varrho}$ requires that its components also resolve holistically within the arbitrarily selected coordinate system. The elimination of this model tensor according to Riemann therefore corresponds to an indispensable prerequisite for the fact that it appears possible to achieve invariability (constancy) of all $g_{\mu v}$ by means of a matching (suitable) selection of the coordinate system. Mathematicians have evaluated the existence of this condition as well as the sufficiency of this condition. In the case of this question, this is similar to the evaluation that the special theory of relativity is valid in the case of a matching selection of the reference system within non-infinite (multi-dimensional time-space) areas.

By reducing equation (43) with regard to the specifications τ and ϱ, the covariant model tensor with rank 2 is generally formed

$$(44) \quad \left\{ \begin{aligned} &B_{\mu v} = R_{\mu v} + S_{\mu v} \\ &R_{\mu v} = -\Omega/\Omega x_\alpha \{\mu\, v \mid \alpha\} + \{\mu\, \alpha \mid \beta\}\{v\, \beta \mid \alpha\} \\ &S_{\mu v} = \Omega\lg\sqrt{-g}/\Omega x_\mu \Omega x_v - \{\mu\, v \mid \alpha\}\Omega\lg\sqrt{-g}/\Omega x_\alpha \end{aligned} \right. \qquad \text{(Einstein, 1916, p. 801).}$$

Side note regarding the point coordinate selection. It was already learned in section 8 after the system of equations (18a) that the point selection can be carried out advantageously so that $\sqrt{-g} = 1$ applies. A consideration of the equations within the two previous sections explains that the design laws of the model tensors experience a

Kolek, Erik (2024). On the physical foundations of interstellar space travel. In: *Chronicles of Business Informatics Physics (CBIP)*. Volume 2, edition no. 1.0. ISBN: 9783759731159.

significantly higher simplicity by means of an appropriate selection. This is particularly true with regard to the model tensor $B_{\mu\nu}$ just derived, which has a fundamental task in this theory to be described. The considered particularity with regard to the point selection causes, so to speak, the resolution of $S_{\mu\nu}$, whereby the model tensor $B_{\mu\nu}$ can be reduced to $R_{\mu\nu}$.

Albert Einstein therefore wanted to note below each relationship in its simplified form of expression that has the special feature of point selection explained. It is then easy to derive the general covariant systems of equations, if this is required in particular.

C. Theory of the gravitational field and its interaction.

§ 13. Moving body system of a point of mass within the gravitational field.

Equation regarding the field components of gravity (= inertia).

According to the special theory of relativity, a body that is not exposed to any external influences can (therefore) move linearly and homogeneously in isolation. The same principle can also be applied according to the general theory of relativity to a region of space with four dimensions (mentally), in which the system of bodies in motion (reference system) K_0 can be selected in accordance with each other and chosen accordingly, so that all $g_{\mu\nu}$ receive their special invariant quantities determined according to (4a).

Albert Einstein briefly looked at the above body motion starting from an arbitrarily chosen reference frame K_1, then {this mass} propagates as seen from there, understood according to the thoughts in section 2 within a gravitational field (which attracts the body). A law of direction of motion with respect to K_1 can simply be derived from the following thought. With respect to K_0, this law resembles a line with four dimensions, i.e. a geodesic straight line (like a bent cube or parallelepiped that is no longer straight and uniform). Because this geodesic straight line is not thought of as dependent on a coordinate system, the corresponding system of equations must also

Kolek, Erik (2024). On the physical foundations of interstellar space travel. In: *Chronicles of Business Informatics Physics (CBIP)*. Volume 2, edition no. 1.0. ISBN: 9783759731159.

represent the system of bodies of motion of a point of mass with respect to K_1. Albert Einstein thought that

(45) $\Gamma_{\mu\nu}{}^{\tau} = -\{\mu\ \nu\ |\ \tau\}$ (Einstein, 1916, p. 802) is,

then the system of equations for the point direction movement with respect to K_1 determines

(46) $d^2x_{\tau}/ds^2 = \Gamma_{\mu\nu}{}^{\tau} \times dx_{\nu}/ds \times dx_{\nu}/ds$ (Einstein, 1916, p. 802).

Albert Einstein now understood an extremely simple theory, in the form of a generally covariant equation regarding the point motion in a gravitational field, which is also given if he (again) imagined away the coordinate system K_0, with respect to which the special theory of relativity has its validity within a non-infinite space. He was able to think this theory accordingly, because (46) only includes the first deductions (constellations) of all $g_{\mu\nu}$, in between there are no derivations (relations), also not in the special case of the existence of the coordinate system K_0. This is because these relations exist from the second (with the first) derivation according to the relations $B_{\mu\sigma\tau}{}^{\varrho} = 0$ imagined in section 12.

If the $\Gamma_{\mu\nu}{}^{\tau}$ dissolve (or if the $\Gamma_{\mu\nu}{}^{\tau}$ move away like a body), then a mass point propagates (in four dimensions again) linearly and homogeneously; the vector values mentioned are the reason why a divergence (deviation) of this direction of movement of a body can arise with regard to its homogeneity. (The general theory of relativity is therefore based on a non-linear and heterogeneous translational motion of bodies influenced by gravity); these are the components of the gravitational field.

Counter-equation with regard to the field components for determining anti-gravity.

Erik Kolek now adds another coordinate system K_2 to the one just mentioned, in which a gravitational field exists as in K_1 and must therefore also have an influence on the geodesic body line movement in four dimensions. As the only observer, he therefore

Kolek, Erik (2024). On the physical foundations of interstellar space travel. In: *Chronicles of Business Informatics Physics (CBIP)*. Volume 2, edition no. 1.0. ISBN: 9783759731159.

no longer only considers an asteroid impact on the earth as in Albert Einstein's general theory of relativity, but also considers the influence of the moon's gravitational field on this reference body movement of the asteroid with a geodesic course towards the earth; the gravitational field of the moon with lower gravity (= inertia) should therefore have an opposing divergent, slowing influence on the body movement speed of the asteroid. On the basis of this completion by the anti-general theory of relativity, he generally assumes

(45a) $-\Gamma_{\mu v}^{\tau} = \{\mu\ v \mid \tau\}$ (Einstein, 1916, p. 802) and

(46a) $d^2x_\tau/ds^2 = -\Gamma_{\mu v}^{\tau} \times dx_v/ds \times dx_v/ds$ (Einstein, 1916, p. 802).

On the basis of this counter-equation with regard to the field components for determining the anti-gravity, the summed equation of the point movement with regard to K_1 (here body impact) and K_2 (here body attraction) results as a consequence.

(45b) $(\Gamma_{\mu v}^{\tau})_1 + (-\Gamma_{\mu v}^{\tau})_2 = -\{\mu\ v \mid \tau\}_1 + \{\mu\ v \mid \tau\}_2$ (Einstein, 1916, p. 802) and

(46b) $(d^2x_\tau/ds^2)_1 + (d^2x_\tau/ds^2)_2 = (\Gamma_{\mu v}^{\tau} \times dx_v/ds \times dx_v/ds)_1 + (-\Gamma_{\mu v}^{\tau} \times dx_v/ds \times dx_v/ds)_2$ (Einstein, 1916, p. 802),

or for the anti-general case in a coinciding point of K_1 and K_2

(46c) $d^2x_\tau/ds^2 = (\Gamma_{\mu v}^{\tau} - \Gamma_{\mu v}^{\tau})(d^2x_v/ds^2)/2$ (Einstein, 1916, p. 802).

If the $\Gamma_{\mu v}^{\tau}$ disappear, or if they are zero, then the moving body will simply remain stationary between the two gravitational fields of K_1 and K_2 (neutral gravitational field zone), but only practically, as it must maintain its rotational motion, according to the overall equation of point motion (46b or 46c), and its translational motion at a certain point on a further geodesic line. Thus, in the anti-general case, moons as well as planets and other celestial objects are captured between two gravitational fields of arbitrarily selected ponderable point masses. If their vector sizes are different, there will also be a non-linear and heterogeneous body line motion, but this will be less accelerated in our physical case due to the second body (here the Moon), which also

Kolek, Erik (2024). On the physical foundations of interstellar space travel. In: *Chronicles of Business Informatics Physics (CBIP)*. Volume 2, edition no. 1.0. ISBN: 9783759731159.

generates an attractive gravitational field, which must lead to a change in the angle, measured in arc seconds, and a slowdown in the speed of the asteroid with respect to the body collision on K_1 (here the Earth). Habitable planets with one or more moons therefore offer a certain degree of protection against matter collisions.

§ 14. The gravitational field equations without the presence of mass.

Albert Einstein subsequently differentiated between the "gravitational field" and the "mass" in such a way that every point in the universe is understood as a "mass", with the exception of the gravitational field, which is not (no longer) understood as a "mass". Accordingly, the concept of mass is not applied according to its usual meaning, since the electromagnetic field component (vector component) of the gravitational field is also not described as a mass.

The subsequent role of Albert Einstein was to derive the gravitational equations of the (remaining) field without the presence of mass. To do this, he again used the same procedure as in the previous section during the deduction of the system of equations for the directional motion of a point of mass. No general case, within which the gravitational equations to be set up obviously have to apply, represents the case of the initial theory of relativity, within which all $g_{\mu\nu}$ have certain invariant quantities. This represents the special case within a certain, non-infinite range (with four dimensions) with respect to a fixed reference system K_0. In this coordinate system, all components $B_{\mu\sigma\tau}{}^{\varrho}$ of the model tensor dissolve (or become zero) according to Riemann (43). The Riemann tensor components then also lose their significance with regard to the observed area and also with regard to all other body line systems (reference systems).

All systems of equations to be derived with regard to the massless gravitational field therefore obviously have a (direct) truth hidden within them as soon as the $B_{\mu\sigma\tau}{}^{\varrho}$ are omitted. However, this relation represents an equation that is described too comprehensively. Since it should be easy to understand that, for example, a gravitational field in the vicinity of a point of matter is in no case "zero-transformable"

Kolek, Erik (2024). On the physical foundations of interstellar space travel. In: *Chronicles of Business Informatics Physics (CBIP)*. Volume 2, edition no. 1.0. ISBN: 9783759731159.

by means of any selection of a reference system, i.e. it is transformable with respect to a special case of constant $g_{\mu\nu}$.

Therefore, the derivation of the elimination of the *non-anti-symmetric* model tensor $B_{\mu\nu}$ from the Riemann tensor $B_{\mu\sigma\tau}{}^{\varrho}$ with respect to the massless gravity field is comprehensible. In general, this method results in 10 systems of equations with regard to the 10 values $g_{\mu\nu}$, which apply to the special case if all $B_{\mu\sigma\tau}{}^{\varrho}$ are omitted. The gravitational equations, taking into account (44) with regard to the selection of the reference system specified by Albert Einstein with regard to the massless field area (with four dimensions), are called

$$
(47) \quad \left\{ \begin{array}{l} \Omega\Gamma_{\mu\nu}{}^{\alpha}/\Omega x_{\alpha} + \Gamma_{\mu\beta}{}^{\alpha}\Gamma_{\nu\alpha}{}^{\beta} = 0 \\[6pt] \sqrt{-g} = 1 \end{array} \right. \qquad \text{(Einstein, 1916, p. 803).}
$$

At this point, the reference must be linked so that this selection of equation systems results in the smallest possible restriction (limitation). Since there is no model tensor with rank 2 in addition to $B_{\mu\nu}$, which can be designed using $g_{\mu\nu}$ and the associated deductions, higher third-degree derivatives are missing and the (set up) second-degree derivatives are straightforward. This minimal arbitrariness in the gravitational field equations (47) can otherwise only be assumed with regard to the model tensor $B_{\mu\nu} + \lambda g_{\mu\nu}(g^{\alpha\beta}B_{\alpha\beta})$, in which λ represents the invariant. However, if this expression is generally set to 0, then the system of equations $B_{\mu\nu} = 0$ is generally inferred (or this formulation is generally found again in reverse).

The gravitational field equations (47) fulfill a condition according to the general theory of relativity, which requires, by means of a completely mathematical method, merging systems of equations connected with the systems of bodies of motion (46), which as a first instance resembles Newton's law of attraction. As a second instance, it explains the reason for the periheliptic motion of the planet Mercury found by Leverrier (left behind after integration of error corrections), which in Albert Einstein's opinion had to be a convincing proof of the physical truth of his theory.

Kolek, Erik (2024). On the physical foundations of interstellar space travel. In: *Chronicles of Business Informatics Physics (CBIP)*. Volume 2, edition no. 1.0. ISBN: 9783759731159.

With regard to Erik Kolek's anti-general theory of relativity, Albert Einstein's gravitational equations (47) remain unchanged, as these are derived with regard to a massless field. This means that even for our example of the Earth and Moon (K_1 and K_2), this fundamental, actually massless (also electromassless) but nevertheless gravity-generating field remains because, according to Albert Einstein's general theory of relativity, the reference bodies must always move within this gravitational field and on geodesic, i.e. non-linear and heterogeneous, lines (orbital curves) in our universe; otherwise Newton, Leverrier and Albert Einstein would be wrong in their theories of body mechanics, but this is ruled out by the confirmed perihelion motion of Mercury alone. As just described, I am of the same opinion as Albert Einstein, which means that the anti-general theory of relativity, as a completion (consequence) of the general theory of relativity, must also correspond to a correctness according to physics.

§ 15. The property determining the gravitational field according to Hamilton, energy momentum theorem.

As an evaluation of the fact that Albert Einstein's gravitational field equations fulfill the energy momentum theorem, it seems to be easiest to formulate the derivation according to the following Hamiltonian expression (again):

$$\delta\{\textstyle\int H d\tau\} = 0$$

$$(47a) \quad \Big\{ \quad H = g^{\mu\nu}\Gamma_{\mu\beta}{}^{\alpha}\Gamma_{\nu\alpha}{}^{\beta}$$

$$\sqrt{-g} = 1 \qquad\qquad \text{(Einstein, 1916, p. 804).}$$

Here, all deviations (divergences, variations) dissolve at all boundaries of the viewed endless integration (time) space with four dimensions; (the gravitational field thus exists constantly everywhere in the universe).

First, it must be proved that the expression (47a) is equivalent (analogous, equal) with respect to the previous notation of the gravitational field equations given in (47). Therefore, Albert Einstein adopted H as a property of all $g_{\mu\nu}$ as well as of all

Kolek, Erik (2024). On the physical foundations of interstellar space travel. In: *Chronicles of Business Informatics Physics (CBIP)*. Volume 2, edition no. 1.0. ISBN: 9783759731159.

$g^{\mu\nu}{}_\sigma = (= \Omega g^{\mu\nu}/\Omega x_\sigma)$ (Einstein, 1916, p. 804).

This corresponds first

$$\delta H = \delta g^{\mu\nu}\Gamma_{\mu\beta}{}^\alpha\Gamma_{\nu\alpha}{}^\beta + 2g^{\mu\nu}\Gamma_{\mu\beta}{}^\alpha\Gamma_{\nu\alpha}{}^\beta$$
$$= -\delta g^{\mu\nu}\Gamma_{\mu\beta}{}^\alpha\Gamma_{\nu\alpha}{}^\beta + \delta(g^{\mu\nu}\Gamma_{\nu\alpha}{}^\beta)2\Gamma_{\mu\beta}{}^\alpha \text{ (Einstein, 1916, p. 804)}.$$

Now, however

$$\delta(g^{\mu\nu}\Gamma_{\nu\alpha}{}^\beta) = -1/2\delta[g^{\mu\nu}g^{\beta\lambda}(\Omega g_{\nu\lambda}/\Omega x_\alpha + \Omega g^{\alpha\lambda}/\Omega x_\nu - \Omega g_{\alpha\nu}/\Omega x_\lambda)] \text{ (Einstein, 1916, p. 804)}.$$

Within the round bracket on the right-hand side, the two members in the last position have different protruding symbols and differ from each other (due to the respective sign) because the notation of the sum specifications is freely selectable, which arise due to the exchange of the specifications of μ and β with λ in each case. These cancel each other out in the equation with regard to δH, as they are multiplied by the value $\Gamma_{\mu\beta}{}^\alpha$, which is *not anti-symmetrical* with regard to the values μ and β. Therefore, only the first term in the round bracket needs to be taken into account so that the following generally follows, taking into account (31)

$$\delta H = -\delta g^{\mu\nu}\Gamma_{\mu\beta}{}^\alpha\Gamma_{\nu\alpha}{}^\beta - \delta g^{\mu\beta}{}_\alpha I^\alpha{}_{\mu\beta} \text{ (Einstein, 1916, p. 805)}.$$

Now therefore resembles

$$\Omega H/\Omega g^{\mu\nu} = -$$

$$(48) \quad \{ \quad \Gamma_{\mu\beta}{}^\alpha\Gamma_{\nu\alpha}{}^\beta$$

$$\Omega H/\Omega g^{\mu\nu}{}_\sigma = \Gamma_{\mu\nu}{}^\sigma \qquad \text{(Einstein, 1916, p. 805)}.$$

After performing a substitution (variation, change) in (47a), the equation

(47b) $\Omega/\Omega x_\alpha(\Omega H/\Omega g^{\mu\nu}{}_\alpha) - \Omega H/\Omega g^{\mu\nu} = 0$ (Einstein, 1916, p. 805),

which is coincident with equation (47) due to (48), which had to be evaluated. – If (47b) is generally multiplied by the $g^{\mu\nu}{}_\sigma$, then the general result is that

$\Omega g^{\mu\nu}{}_\sigma/\Omega x_\alpha = \Omega g^{\mu\nu}{}_\alpha/\Omega x_\sigma$ (Einstein, 1916, p. 805)

Kolek, Erik (2024). On the physical foundations of interstellar space travel. In: *Chronicles of Business Informatics Physics (CBIP)*. Volume 2, edition no. 1.0. ISBN: 9783759731159.

and the following

$$g^{\mu\nu}{}_\sigma \times \Omega/\Omega x_\alpha(\Omega H/\Omega g^{\mu\nu}{}_\alpha) = \Omega/\Omega x_\alpha(g^{\mu\nu}{}_\sigma \times \Omega H/\Omega g^{\mu\nu}{}_\alpha) - \Omega H/\Omega g^{\mu\nu}{}_\alpha \times \Omega g^{\mu\nu}{}_\alpha/\Omega x_\alpha$$

(Einstein, 1916, p. 805)

the system of equations

$$\Omega/\Omega x_\alpha(g^{\mu\nu}{}_\sigma \times \Omega H/\Omega g^{\mu\nu}{}_\alpha) - \Omega H/\Omega x_\sigma = 0 \text{ (Einstein, 1916, p. 805)}$$

respectively (in the equation $-2\varkappa$ is introduced as a factor, the reason for this will be explained later).

$$(49) \quad \left\{ \begin{array}{l} \Omega t^\alpha{}_\sigma/\Omega x_\alpha = 0 \\ -2\varkappa t^\alpha{}_\sigma = g^{\mu\nu}{}_\sigma \times \Omega H/\Omega g^{\mu\nu}{}_\alpha - \delta^\alpha{}_\sigma H \end{array} \right. \quad \text{(Einstein, 1916, p. 805),}$$

or, due to (48), the system of equations in the second position of (47) and (34)

$$(50)\ \varkappa t^\alpha{}_\sigma = 1/2 \times \delta^\alpha{}_\sigma g^{\mu\nu}\Gamma_{\mu\beta}{}^\alpha\Gamma_{\nu\alpha}{}^\beta - g^{\mu\nu}\Gamma_{\mu\beta}{}^\alpha\Gamma_{\nu\sigma}{}^\beta \text{ (Einstein, 1916, p. 806).}$$

It is important to learn that $t^\alpha{}_\sigma$ does not represent a model tensor; instead, (49) is valid with regard to any reference system in which $\sqrt{-g} = 1$ applies. With regard to the gravitational field, the system of equations describes the associated law of conservation of energy and law of conservation of momentum. In fact, merging (integrating) this system of equations using a volume V with three dimensions produces the following four systems of equations

$$(49a)\ d/dx_4\left\{\int t_\sigma^4 dV\right\} = \int(t_\sigma^1\alpha_1 + t_\sigma^2\alpha_2 + t_\sigma^3\alpha_3)dS \text{ (Einstein, 1916, p. 806),}$$

in which α_1 to α_3 indicate the cosine of the direction of an internally curved straight line with respect to the boundary of the surface element (model vector) with the value *dS* (according to Euclid's geometry). In general, the equation regarding the axioms of conservation can be found in the same representation. Albert Einstein called all values t_σ^α "energy components" of the gravitational field.

Kolek, Erik (2024). On the physical foundations of interstellar space travel. In: *Chronicles of Business Informatics Physics (CBIP)*. Volume 2, edition no. 1.0. ISBN: 9783759731159.

Albert Einstein now wanted to write down the system of equations (47) as a third derivative, which greatly supports a true-to-life understanding for him with regard to the object under consideration (gravitational field). If the gravitational equations (47) are multiplied by the $g^{v\sigma}$, they are generally transformed into a "mixed" expression. The following must generally be taken into account

$$g^{v\sigma} \times \Omega\Gamma_{\mu v}{}^{\alpha}/\Omega x_{\alpha} = \Omega/\Omega x_{\alpha} \times (g^{v\sigma}\Gamma_{\mu v}{}^{\alpha}) - \Omega g^{v\sigma}/\Omega x_{\alpha} \times \Gamma_{\mu v}{}^{\alpha} \text{ (Einstein, 1916, p. 806),}$$

whose value due to (34) is analogous to

$$\Omega/\Omega x_{\alpha}(g^{v\sigma}\Gamma_{\mu v}{}^{\alpha}) - g^{v\beta}\Gamma_{\alpha\beta}{}^{\sigma}\Gamma_{\mu v}{}^{\alpha} - g^{\sigma\beta}\Gamma_{\beta\alpha}{}^{v}\Gamma_{\mu v}{}^{\alpha} \text{ (Einstein, 1916, p. 806),}$$

or (subsequently as soon as the designation of the totals has been changed) is analogous to

$$\Omega/\Omega x_{\alpha}(g^{\sigma\beta}\Gamma_{\mu\beta}{}^{\alpha}) - g^{mn}\Gamma_{m\beta}{}^{\sigma}\Gamma_{n\mu}{}^{\beta} - g^{v\sigma}\Gamma_{\mu\beta}{}^{\alpha}\Gamma_{v\alpha}{}^{\beta} \text{ (Einstein, 1916, p. 806).}$$

The last term in this equation is solved by the term that appears, which is created by the term that is in the second position within the gravitational equations (47); for the exchange of this second term of the same set, a relation to (50) can be set ($t = t_{\alpha}{}^{\alpha}$).

$$\varkappa(t_{\mu}{}^{\sigma} - 1/2 \times \delta_{\mu}{}^{\sigma}t) \text{ (Einstein, 1916, p. 806)}$$

In general, therefore, instead of the system of equations (47)

$$(51) \quad \left\{ \begin{array}{c} \Omega/\Omega x_{\alpha}(g^{\sigma\beta}\Gamma_{\mu\beta}{}^{\alpha}) = -\varkappa(t_{\mu}{}^{\sigma} - 1/2 \times \delta_{\mu}{}^{\sigma}t) \\ \sqrt{-g} = 1 \end{array} \right. \quad \text{(Einstein, 1916, p. 806).}$$

For the *anti-general case* in which (at least) two (or more) gravitational fields can influence each other, the addition result applies analogously to (51)

$$(51a) \quad \left\{ \begin{array}{c} 2[\Omega/\Omega x_{\alpha}(g^{\sigma\beta}\Gamma_{\mu\beta}{}^{\alpha})] = -2\varkappa(t_{\mu}{}^{\sigma} - 1/2 \times \delta_{\mu}{}^{\sigma}t) \\ 2\sqrt{-g} = 2 \end{array} \right. \quad \text{(Einstein, 1916, p. 806).}$$

According to (51a), the number of gravitational fields has no influence on the fulfillment of the energy momentum theorem by the gravitational field equations of

Kolek, Erik (2024). On the physical foundations of interstellar space travel. In: *Chronicles of Business Informatics Physics (CBIP)*. Volume 2, edition no. 1.0. ISBN: 9783759731159.

Albert Einstein (51); the energy momentum theorem is therefore also fulfilled between (at least) two (or more) gravitational fields in the *anti-general theory of relativity*.

Of course, in the *anti-general case*, the instantaneous energy momentum theorem can depend on gravitational fields of different strengths or have different effects and still always be fulfilled.

$$(51b) \quad \left\{ \begin{array}{c} \Omega/\Omega x_\alpha(g^{\sigma\beta}\Gamma_{\mu\beta}{}^\alpha) \neq \Omega/\Omega x_\alpha(g^{\sigma\beta}\Gamma_{\mu\beta}{}^\alpha) \\[4pt] -\varkappa(t_\mu{}^\sigma - 1/2 \times \delta_\mu{}^\sigma t) \neq -\varkappa(t_\mu{}^\sigma - 1/2 \times \\[4pt] \delta_\mu{}^\sigma t) \\[4pt] \sqrt{-g} = 1 \end{array} \right. \qquad \text{(Einstein, 1916, p. 806).}$$

§ 16. Generalized versions of the gravitational field equations.

All derived gravitational field equations from the last section with validity for massless space must agree with Newton's gravitational theory equation

$\Delta\varphi = 0$ (Einstein, 1916, p. 807)

can be compared. Albert Einstein had to find the field equations that correspond to Poisson's system of equations

$\Delta\varphi = 4\pi\varkappa\varrho$ (Einstein, 1916, p. 807)

because here ϱ denotes the mass density.

The result from the special theory of relativity, i.e. that the mass inertia is equal to the mass energy, has its perfect mathematical representation in a *non-anti-symmetric* model tensor with rank 2, an energy model tensor. Albert Einstein therefore also introduced the energy model tensor for the mass $T_\sigma{}^\alpha$ within the general theory of relativity, which should have a mixed property equal to the energy model components $t_\sigma{}^\alpha$ [equation systems (49) and (50)] of a gravitational field, but is assigned to a *non-anti-symmetric* covariant model tensor. Selected *non-anti-symmetric* model tensors are $g_{\sigma\tau}T_\sigma{}^\alpha = T_{\sigma\tau}$ and $g^{\sigma\beta}T_\sigma{}^\alpha = T^{\alpha\beta}$.

Kolek, Erik (2024). On the physical foundations of interstellar space travel. In: *Chronicles of Business Informatics Physics (CBIP)*. Volume 2, edition no. 1.0. ISBN: 9783759731159.

The only sensible integration of the energy model tensor (corresponding to the matter density ϱ from Poisson's system of equations) into all gravitational field equations is given by equation (51). If, for example, a complete reference system (such as our solar system) is generally investigated, then the total system mass, and therefore also the total gravitational effect, must be dependent on the total system energy, i.e. on the mass energy and gravitational field energy together. This can be described by the fact that generally within (51), instead of the energy model components $t_\mu{}^\sigma$ of a gravitational field, only the summations $t_\mu{}^\sigma + T_\mu{}^\sigma$ of all energy model components are integrated with regard to the matter field and gravitational field. In general, the following model tensor equation is obtained instead of (51)

$$\Omega/\Omega x_\alpha(g^{\sigma\beta}\Gamma_\mu{}^\alpha{}_\beta) = -\varkappa[(t_\mu{}^\sigma + T_\mu{}^\sigma) - 1/2 \times$$

(52) $\{$ $\qquad\qquad\qquad \delta_\mu{}^\sigma(t + T)]$

$$\sqrt{-g} = 1 \qquad\qquad \text{(Einstein, 1916, p. 807)},$$

where $T = T_\mu{}^\mu$ is assumed (invariant according to Laue). These correspond to the general gravitational field equations with a mixed character. Instead of (47), the following system of equations is obtained retrospectively

$$\Omega\Gamma_\mu{}^\alpha{}_\nu/\Omega x_\alpha + \Gamma_\mu{}^\alpha{}_\beta\Gamma_\nu{}^\beta{}_\alpha = -\varkappa(T_{\mu\nu} - 1/2$$

(53) $\{$ $\qquad\qquad\qquad \times g_{\mu\nu}T)$

$$\sqrt{-g} = 1 \qquad\qquad \text{(Einstein, 1916, p. 808)}.$$

A limitation, however, is that the previous integration of the energy model tensor for the mass only by means of the principle of relativity has no (sufficient) foundation; therefore, Albert Einstein had to establish {prove} this beforehand on the basis of the condition that the field energy of gravitation gravitates uniformly, just like all other types of field energies (for example of matter). However, the most important reason for choosing the previous systems of equations is that, as a consequence, the momentum and energy conservation equations are valid with respect to the components of the total energy, which correspond exactly to the systems of equations (49) and (49a). This is described in the next section.

Kolek, Erik (2024). On the physical foundations of interstellar space travel. In: *Chronicles of Business Informatics Physics (CBIP)*. Volume 2, edition no. 1.0. ISBN: 9783759731159.

From two or more gravitational fields, the following therefore applies according to the *anti-general theory of relativity*, analogous to the explanations in the previous and section 16

$$\Omega\Gamma_\mu{}^\alpha{}_\nu/\Omega x_\alpha + \Gamma_\mu{}^\alpha{}_\beta\Gamma_\nu{}^\beta{}_\alpha \neq \Omega\Gamma_\mu{}^\alpha{}_\nu/\Omega x_\alpha + \Gamma_\mu{}^\alpha{}_\beta\Gamma_\nu{}^\beta{}_\alpha$$

(53a) $\quad\{\quad -\varkappa(T_{\mu\nu} - 1/2 \times g_{\mu\nu}T) \neq -\varkappa(T_{\mu\nu} - 1/2 \times g_{\mu\nu}T)$

$$\sqrt{-g} = 1 \qquad\qquad \text{(Einstein, 1916, p. 808)}.$$

§ 17. The conservation equations for the general case.

The system of equations (52) can simply be rearranged accordingly so that the second term is resolved with respect to the left-hand side. To do this, (52) must be shortened with regard to the values μ and σ and the result obtained by multiplying $1/2\delta_\mu{}^\sigma$ by the system of equations in (52) must be subtracted. From this follows

(52a) $\quad \Omega/\Omega x_\alpha(g^{\sigma\beta}\Gamma_\mu{}^\alpha{}_\beta - 1/2 \times \delta_\mu{}^\sigma g^{\lambda\beta}\Gamma_\lambda{}^\alpha{}_\beta) = -\varkappa(t_\mu{}^\sigma + T_\mu{}^\sigma)$ (Einstein, 1916, p. 808).

Albert Einstein applied the procedure $\Omega/\Omega x_\sigma$ to this system of equations. Now follows

$$\Omega^2/\Omega x_\alpha\Omega x_\sigma(g^{\sigma\beta}\Gamma_\mu{}^\alpha{}_\beta)$$

$$= -1/2 \times \Omega^2/\Omega x_\alpha\Omega x_\sigma[g^{\sigma\beta}g^{\alpha\lambda}(\Omega g_{\mu\lambda}/\Omega x_\beta + \Omega g_{\beta\lambda}/\Omega x_\mu - \Omega g_{\mu\beta}/\Omega x_\lambda)] \text{ (Einstein, 1916, p. 808)}.$$

The first and third expressions in the round brackets produce results that cancel each other out. This becomes apparent as soon as all summation specifications α and σ on one side and β and λ on the other side are exchanged within the result in the third expression (in the square bracket). The expression on the right-hand side can be transformed according to (31), resulting in

(54) $\Omega^2/\Omega x_\alpha\Omega x_\sigma(g^{\sigma\beta}\Gamma_\mu{}^\alpha{}_\beta) = 1/2 \times \Omega^3 g^{\alpha\beta}/\Omega x_\alpha\Omega x_\beta\Omega x_\mu$ (Einstein, 1916, p. 808).

Kolek, Erik (2024). On the physical foundations of interstellar space travel. In: *Chronicles of Business Informatics Physics (CBIP)*. Volume 2, edition no. 1.0. ISBN: 9783759731159.

The expression in the second position opposite the right-hand side of (52a) temporarily results in

$$- 1/2 \times \Omega^2/\Omega x_\alpha \Omega x_\mu (g^{\lambda\beta}\Gamma_\lambda{}^\alpha{}_\beta) \text{ (Einstein, 1916, p. 808)}$$

respectively

$$1/4 \times \Omega^2/\Omega x_\alpha \Omega x_\mu [g^{\lambda\beta}g^{\alpha\delta}(\Omega g_{\delta\lambda}/\Omega x_\beta + \Omega g_{\delta\beta}/\Omega x_\lambda - \Omega g_{\lambda\beta}/\Omega x_\delta)] \text{ (Einstein, 1916, p. 809)}.$$

The expression originating from the last-placed expression in the round bracket dissolves due to (29) during a coordinate point selection made by Albert Einstein. All two remaining members can be summarized and bring together due to (31)

$$- 1/2 \times \Omega^3 g^{\alpha\beta}/\Omega x_\alpha \Omega x_\beta \Omega x_\mu \text{ (Einstein, 1916, p. 809)},$$

taking into account (54) the following analogy

$$(55) \ \Omega^2/\Omega x_\alpha \Omega x_\sigma (g^{\sigma\beta}\Gamma_\mu{}^\alpha{}_\beta - 1/2 \times \delta_\mu{}^\sigma g^{\lambda\beta}\Gamma_\lambda{}^\alpha{}_\beta) \equiv 0 \text{ (Einstein, 1916, p. 809)}.$$

With the aid of (55) and (52a), the sequence

$$(56) \ \Omega(t_\mu{}^\sigma + T_\mu{}^\sigma)/\Omega x_\sigma = 0 \text{ (Einstein, 1916, p. 809)}.$$

Using Albert Einstein's gravitational field equations, it can therefore be recognized that these correspond to the law of conservation of momentum and energy. In general, this can be seen most easily with the investigation that leads to the system of equations (49a); here, instead of all energy model components $t_\mu{}^\sigma$ of a gravitational field, the total energy components with regard to the mass field and gravitational field are to be established. This generally results in

$$(49b) \ d/dx_4\{\textstyle\int T_\sigma{}^4 dV\} = \textstyle\int (T_\sigma{}^1\alpha_1 + T_\sigma{}^2\alpha_2 + T_\sigma{}^3\alpha_3)dS \text{ (Einstein, 1916, p. 806)}.$$

The same as described above also applies to the *anti-general case*, because here too the law of conservation of momentum and energy is observed on the basis of the gravitational field equations.

Kolek, Erik (2024). On the physical foundations of interstellar space travel. In: *Chronicles of Business Informatics Physics (CBIP)*. Volume 2, edition no. 1.0. ISBN: 9783759731159.

(56a) $- \Omega(t_\mu{}^\sigma + T_\mu{}^\sigma)/\Omega x_\sigma = 0$ (Einstein, 1916, p. 809).

§ 18. The consequence of the gravitational field equations is similar to the momentum energy axiom with regard to mass.

If (53) is multiplied by $\Omega g^{\mu\nu}/\Omega x_\sigma$, the method used in section 15, taking into account the resolution in terms of

$g_{\mu\nu}(\Omega g^{\mu\nu}/\Omega x_\sigma)$ (Einstein, 1916, p. 809)

the system of equations

$\Omega t_\sigma{}^\alpha/\Omega x_\alpha + 1/2 \times (\Omega g^{\mu\nu}/\Omega x_\sigma)T_{\mu\nu} = 0$ (Einstein, 1916, p. 809),

or taking into account with regard to (56)

(57) $\Omega T_\sigma{}^\alpha/\Omega x_\alpha + 1/2 \times (\Omega g^{\mu\nu}/\Omega x_\sigma)T_{\mu\nu} = 0$ (Einstein, 1916, p. 809).

The comparison of (57) and (41b) illustrates that this system of equations resembles the resolution of the tensor deviation of the mass-energy components due to the selection made with regard to the coordinate reference system. According to this physics, a realization of the second-placed term with respect to the right-hand side of the equation means that, with respect to the mass, only the momentum and energy conservation axioms are invalid in the strict sense, or are valid only in the case where the $g^{\mu\nu}$ appear uniform, i.e. as soon as the strength of the gravitational field becomes zero. The second-placed term resembles a reference system in terms of energy or momentum, which acts on the mass with every time bar and every change in volume, starting from the gravitational field. This becomes even easier to understand as soon as (57) is replaced by (41) in general

(57a) $\Omega T_\sigma{}^\alpha/\Omega x_\alpha = - \Gamma_\sigma{}^\alpha{}_\beta T_\alpha{}^\beta$ (Einstein, 1916, p. 810).

The right-positioned link is to be understood as the energy-based (equal to momentum-based) influence of the gravitational field with regard to the mass.

Kolek, Erik (2024). On the physical foundations of interstellar space travel. In: *Chronicles of Business Informatics Physics (CBIP)*. Volume 2, edition no. 1.0. ISBN: 9783759731159.

All gravitational field equations therefore contain four requirements at the same time, which the events of mass must fulfill. These make it possible to complete all systems of equations of our mass action as soon as the action can be described by means of four systems of differential equations that do not depend on it. See (Einstein, 1916, p. 810): "D. Hilbert, Nachr. d. K. Gesellsch. d. Wiss. zu Göttingen, Math.-phys. Class. p. 3. 1915."

Accordingly, a general transformation to (57b) can be made to express the following according to the basis of *anti-general relativity theory*: The left-placed member is to be understood as the energy-based (equal to momentum-based) influence of the gravitational field with respect to mass.

(57b) $\Gamma_\sigma{}^\alpha{}_\beta T_\alpha{}^\beta = - \Omega T_\sigma{}^\alpha / \Omega x_\alpha$ (Einstein, 1916, p. 810).

D. The "mass-based" event.

All mathematics-based tools designed in B enabled Albert Einstein to directly generalize all physics-based laws of nature regarding mass (water dynamics, electrodynamics according to Maxwell), as expressed within the special theory of relativity, so that they can be inserted into the *(anti-)general theory of relativity*. Here, of course, the *(anti-)general principle of relativity* does not prescribe any additional limitation of all paths; however, this makes the effect of the gravitational field on every event exactly comprehensible, neglecting the need to add any further assumption.

According to this constellation, it is possible that no mandatory conditions need to be added with regard to the physics-based laws of nature of mass (in the narrower sense). Above all, the question of whether both field theories of electromagnetism and gravity together provide a satisfactory basis for this theory of mass or exclude it may remain unanswered. The *(anti-)general principle of relativity* did not allow Albert Einstein to gain any insight into this concept. Therefore, during the extension of the theory, it must be checked whether electromagnetism and gravitational magnetism together are

Kolek, Erik (2024). On the physical foundations of interstellar space travel. In: *Chronicles of Business Informatics Physics (CBIP)*. Volume 2, edition no. 1.0. ISBN: 9783759731159.

able to maintain their performance potential, which is in no way achievable with electromagnetism separately {unthinkable}.

§ 19. Systems of equations according to Euler with regard to frictionless adiabatic liquid materials.

The two invariants should be ρ and ϱ, from which Albert Einstein named the first as "matter pressure" and the last as "matter density" of liquid matter; a system of equations should exist between them (which describes the matter pressure density). A contravariant *non-anti-symmetric* model tensor

(58) $T^{\alpha\beta} = -g^{\alpha\beta}\rho + \varrho(dx_\alpha/ds \times dx_\beta/ds)$ (Einstein, 1916, p. 811)

is intended to represent the contravariant energy model tensor of liquid matter. A covariant model tensor is assigned to the liquid matter energy

(58a) $T_{\mu\nu} = -g_{\mu\nu}\rho + g_{\mu\alpha}(dx_\alpha/ds)g_{\mu\beta}(dx_\beta/ds)\varrho$ (Einstein, 1916, p. 811),

and a mixed model tensor (58b) – for the observer moving with the liquid matter, who imagines a coordinate system in the infinite minimum according to the special theory of relativity, the matter energy density appears to be $T_4^4 = \varrho - \rho$ (or *anti-generally* reformulated $\rho - \varrho = -T_4^4$). This is the fixed understanding with regard to ϱ. Accordingly, ϱ (or ρ) appears to be different with regard to non-compressible (compressible) liquid matter.

(58b) $T_\sigma^\alpha = -\delta_\sigma^\alpha\rho + g_{\sigma\beta}(dx_\beta/ds \times dx_\alpha/ds)\varrho$ (Einstein, 1916, p. 811).

If the right-hand element from (58b) is generally inserted into equation (57a), then the liquid-matter dynamic equation systems of the *(anti-)general theory of relativity* are generally set up according to Euler. The equations cause the compression problem (mechanical difficulty) to disappear without exception; since all four systems of equations (57a) are solved together using the system equation assumed between ϱ and ρ and the system of equations

Kolek, Erik (2024). On the physical foundations of interstellar space travel. In: *Chronicles of Business Informatics Physics (CBIP)*. Volume 2, edition no. 1.0. ISBN: 9783759731159.

$g_{\alpha\beta}(dx_\alpha/ds \times dx_\beta/ds) = 1$ (Einstein, 1916, p. 811)

is sufficient if $g_{\alpha\beta}$ is known for the definition of 6 unknown variables

ρ, ϱ, dx_i/ds ($i = 1, 2, 3, 4$) (Einstein, 1916, p. 811).

If all $g_{\mu\nu}$ are also unknown, then all systems of equations (53) are also included. These correspond to 11 system equations for the determination of all 10 properties $g_{\mu\nu}$, so that the properties $g_{\mu\nu}$ are more than sufficiently fixed. During this determination, it must be taken into account that all systems of equations (57a) within the systems (53) already appear to be taken into account, so that these (53) alone represent a total of 7 non-dependent systems. The latter lack of determinacy (solvability) has the corresponding corroborating justification herein, so that a far-reaching impartiality in the selection of all point coordinates causes the following, whereby this solution difficulty persists mathematically based in a high dimensioning {far-reaching expression} (or this indeterminacy continues to exist), so that three of the spatial properties remain arbitrarily selectable. If no coordinate selection takes place after –g = 1, four spatial properties are still available for arbitrary selection, corresponding to any four properties that can generally be chosen arbitrarily during a coordinate selection.

From the previous statements, the consequence for the contravariant symmetric model tensor based on the hydrodynamic equation systems according to Euler with regard to the *(anti-)general theory of relativity* is as follows

(58c) $\varrho - \rho = T^{\alpha\beta}/g^{\alpha\beta}$ bzw. $\rho - \varrho = - T^{\alpha\beta}/g^{\alpha\beta}$ (Einstein, 1916, p. 811).

The covariant model tensor associated with the contravariant energy density of liquid matter is

(58d) $\varrho - \rho = T_{\mu\nu}/g_{\mu\nu}$ bzw. $\rho - \varrho = - T_{\mu\nu}/g_{\mu\nu}$ (Einstein, 1916, p. 811),

as well as the mixed model tensor

Kolek, Erik (2024). On the physical foundations of interstellar space travel. In: *Chronicles of Business Informatics Physics (CBIP)*. Volume 2, edition no. 1.0. ISBN: 9783759731159.

(58e) $\varrho - \rho = T_\sigma{}^\alpha/\delta_\sigma{}^\alpha$ bzw. $\rho - \varrho = - T_\sigma{}^\alpha/\delta_\sigma{}^\alpha$ (Einstein, 1916, p. 811).

§ 20. Electromagnetic equations of the vacuum field according to Maxwell.

The φ_ν should be the components that describe the covariant four-vector of the power potential of electromagnetism. Using these components, Albert Einstein formed all components $F_{\varrho\sigma}$ of a covariant six-model vector for the electromagnetic field according to (36) in accordance with the equation

(59) $F_{\varrho\sigma} = \Omega\varphi_\varrho/\Omega\varkappa_\sigma - \Omega\varphi_\sigma/\Omega\varkappa_\varrho$ (Einstein, 1916, p. 812).

The consequence of (59) is that the equation

(60) $\Omega F_{\varrho\sigma}/\Omega\varkappa_\tau + \Omega F_{\sigma\tau}/\Omega\varkappa_\varrho + \Omega F_{\tau\varrho}/\Omega\varkappa_\varrho = 0$ (Einstein, 1916, p. 812)

(60 corrected) $\Omega F_{\varrho\sigma}/\Omega\varkappa_\tau + \Omega F_{\sigma\tau}/\Omega\varkappa_\varrho + \Omega F_{\tau\varrho}/\Omega\varkappa_\sigma = 0$ (Einstein, 1916, p. 812)

whose left-placed members according to (37) represent an *anti-symmetric* model tensor with rank 3. A total of 4 systems of equations are therefore assigned to the system of equations (60), which are formulated as follows:

$$
\begin{array}{lll}
& \Omega F_{23}/\Omega\varkappa_4 + \Omega F_{34}/\Omega\varkappa_2 + \Omega F_{42}/\Omega\varkappa_3 = 0 & \text{(Einstein, 1916, p. 812)} \\
& \Omega F_{34}/\Omega\varkappa_1 + \Omega F_{41}/\Omega\varkappa_3 + \Omega F_{13}/\Omega\varkappa_4 = 0 & \text{(Einstein, 1916, p. 812)} \\
\text{(60a)} \quad \{ & \Omega F_{41}/\Omega\varkappa_2 + \Omega F_{12}/\Omega\varkappa_4 + \Omega F_{24}/\Omega\varkappa_1 = 0 & \text{(Einstein, 1916, p. 812)} \\
& \Omega F_{12}/\Omega\varkappa_3 + \Omega F_{23}/\Omega\varkappa_1 + \Omega F_{31}/\Omega\varkappa_2 = 0 & \text{(Einstein, 1916, p. 812).}
\end{array}
$$

A model-imaginative alternative to the previous equation system is:

$$
\begin{array}{lll}
& \Omega F_{14}/\Omega\varkappa_4 + \Omega F_{42}/\Omega\varkappa_2 + \Omega F_{23}/\Omega\varkappa_3 = 0 & \text{(Einstein, 1916, p. 812)} \\
& \Omega F_{24}/\Omega\varkappa_1 + \Omega F_{43}/\Omega\varkappa_3 + \Omega F_{31}\Omega\varkappa_4 = 0 & \text{(Einstein, 1916, p. 812)} \\
\text{(60aa)} \quad \{ & \Omega F_{34}/\Omega\varkappa_2 + \Omega F_{41}/\Omega\varkappa_4 + \Omega F_{12}/\Omega\varkappa_1 = 0 & \text{(Einstein, 1916, p. 812)} \\
& \Omega F_{31}/\Omega\varkappa_3 + \Omega F_{12}/\Omega\varkappa_1 + \Omega F_{23}/\Omega\varkappa_2 = 0 & \text{(Einstein, 1916, p. 812).}
\end{array}
$$

Kolek, Erik (2024). On the physical foundations of interstellar space travel. In: *Chronicles of Business Informatics Physics (CBIP)*. Volume 2, edition no. 1.0. ISBN: 9783759731159.

These systems of equations (60a and 60aa) are each similar to Maxwell's second equation. In general, this becomes immediately apparent as soon as it is said in general terms

$$
\text{(61)} \quad \left\{
\begin{array}{ll}
F_{23} = h_x & F_{14} = e_x \\
F_{31} = h_y & F_{24} = e_y \\
F_{12} = h_z & F_{34} = e_z
\end{array}
\right.
$$

           (Einstein, 1916, p. 813)

           (Einstein, 1916, p. 813)

           (Einstein, 1916, p. 813).

Instead of (60a or 60aa), it is possible to use the usual expression for three-dimensional vector investigations {note}

$$
\text{(60b)} \quad \left\{
\begin{array}{l}
\Omega h / \Omega t + \text{rot } e = 0 \\
\text{div } h = 0
\end{array}
\right.
$$

           (Einstein, 1916, p. 813)

           (Einstein, 1916, p. 813).

Albert Einstein recognized the first reference system after Maxwell by generalizing the system established by Minkowski. He thought of a contravariant six-model vector assigned with respect to $F_{\alpha\beta}$

(62) $F^{\mu\nu} = g^{\mu\alpha} q^{\nu\beta} F_{\alpha\beta}$ (Einstein, 1916, p. 813).

and a contravariant four-model vector J^{μ} for the electric flux density in vacuum; it is then generally possible to apply a scalar equation reference system taking into account (40) oppositely arbitrary exchange possibilities (substitutions) with respect to a determinant 1 (in accordance with the point coordinate selection made by Albert Einstein):

(63) $\Omega F^{\mu\nu} / \Omega x_\nu = J^{\mu}$ (Einstein, 1916, p. 813).

For example, if the following is generally defined

$$
\text{(64)} \quad \left\{
\begin{array}{ll}
F^{23} = h_x{}' & F^{14} = - e_x{}' \\
F^{31} = h_y{}' & F^{24} = - e_y{}' \\
F^{12} = h_z{}' & F^{34} = - e_z{}'
\end{array}
\right.
$$

           (Einstein, 1916, p. 813)

           (Einstein, 1916, p. 813)

           (Einstein, 1916, p. 813),

Kolek, Erik (2024). On the physical foundations of interstellar space travel. In: *Chronicles of Business Informatics Physics (CBIP)*. Volume 2, edition no. 1.0. ISBN: 9783759731159.

the values for the special case of the special {lived initial} theory of relativity appear to be equivalent to all values h_x ... e_z, and additionally

$$J^1 = i_x,\ J^2 = i_y,\ J^3 = i_z,\ J^4 = \varrho \text{ (Einstein, 1916, p. 813),}$$

then generally instead of (63)

$$
\begin{array}{lll}
(63a) \quad \Big\{ & \text{rot } h' - \Omega e'/\Omega t = i & \text{(Einstein, 1916, p. 813)} \\
& \text{div } e' = \varrho & \text{(Einstein, 1916, p. 813).}
\end{array}
$$

Three systems of equations (60, 62 and 63) therefore correspond in general form to Maxwell's vacuum field equations due to the decision made by Albert Einstein regarding the selection of point coordinates.

All energy components of the electromagnetic field. Albert Einstein determined an internal multiplication result (product)

$$(65)\ x_\sigma = F_{\sigma\mu}J^\mu \text{ (Einstein, 1916, p. 814).}$$

According to (61), the corresponding components are three-dimensional shapes

$$
\begin{array}{lll}
& x_1 = \varrho e_x + [i, h]_x & \text{(Einstein, 1916, p. 814)} \\
& \cdots\cdots\cdots\cdots & \text{(Einstein, 1916, p. 814)} \\
(65a) \quad \Big\{ & \cdots\cdots\cdots\cdots & \text{(Einstein, 1916, p. 814)} \\
& x_4 = -(i, e) & \text{(Einstein, 1916, p. 814).}
\end{array}
$$

The covariant four-model vector is x_σ, whose components appear to be equivalent to the negative energy or momentum that is transferred from the (entire) electromagnetic mass to the electromagnetic field for each volume and time scale. If the electric masses appear to be unaffected, i.e. only exposed to the effect of the electromagnetic field, then a covariant four-model vector x_σ must resolve.

To determine the energy components $T_\sigma{}^\nu$ of the electromagnetic field, Albert Einstein only needed to follow the structure of the system of equations (57) for the system of equations $x_\sigma = 0$. With the help of (63) and (65), he first obtained

Kolek, Erik (2024). On the physical foundations of interstellar space travel. In: *Chronicles of Business Informatics Physics (CBIP)*. Volume 2, edition no. 1.0. ISBN: 9783759731159.

$x_\sigma = F_{\sigma\mu}(\Omega F^{\mu\nu}/\Omega x_\nu) = \Omega/\Omega x_\nu(F_{\sigma\mu}F^{\mu\nu}) - F^{\mu\nu}(\Omega F_{\sigma\mu}/\Omega x_\nu)$ (Einstein, 1916, p. 814).

The last-placed expression allows a rearrangement due to (60)

$F^{\mu\nu}(\Omega F_{\sigma\mu}/\Omega x_\nu) = -1/2 \times F^{\mu\nu}(\Omega F_{\mu\nu}/\Omega x_\sigma) = -1/2 \times g^{\mu\alpha}g^{\nu\beta}F_{\alpha\beta}(\Omega F_{\mu\nu}/\Omega x_\sigma)$ (Einstein, 1916, p. 814),

whose last placed member also has the shape

$-1/4[g^{\mu\alpha}g^{\nu\beta}F_{\alpha\beta}(\Omega F_{\mu\nu}/\Omega x_\sigma) + g^{\mu\alpha}g^{\nu\beta}(\Omega F_{\alpha\beta}/\Omega x_\sigma)F_{\mu\nu}]$ (Einstein, 1916, p. 814)

may be assumed. Due to the symmetry, however, the following can be formulated

$-1/4 \times \Omega/\Omega x_\sigma(g^{\mu\alpha}g^{\nu\beta}F_{\alpha\beta}F_{\mu\nu}) + 1/4 \times F_{\alpha\beta}F_{\mu\nu}(\Omega/\Omega x_\sigma)(g^{\mu\alpha}g^{\nu\beta})$ (Einstein, 1916, p. 814).

The very first of the expressions is called in summarized form

$-1/4 \times \Omega/\Omega x_\sigma(F^{\mu\nu}F_{\mu\nu})$ (Einstein, 1916, p. 814),

the second-placed (expression) is created by calculating a derivation and some rearrangement

$-1/2 \times F^{\mu\nu}F_{\mu\nu}g^{\nu\varrho}(\Omega g_{\sigma\tau}/\Omega x_\sigma)$ (Einstein, 1916, p. 815).

If the three expressions obtained are added together, the following relationship generally arises

(66) $x_\sigma = \Omega T_\sigma{}^\nu/\Omega x_\nu - 1/2 \times g^{\tau\mu}(\Omega g_{\mu\nu}/\Omega x_\sigma)T_\tau{}^\nu$ (Einstein, 1916, p. 815),

although

(66a) $T_\sigma{}^\nu = -F_{\sigma\alpha}F^{\nu\alpha} + 1/4 \times \delta_\sigma{}^\nu F_{\alpha\beta}F^{\alpha\beta}$ (Einstein, 1916, p. 815).

The system of equations (66) appears identical with regard to x_σ approaching zero due to (30) and (57) or (57a). All $T_\sigma{}^\nu$ therefore correspond to the energy components of an electromagnetic field. Using (61) and (64), it is generally easy to prove that the energy

Kolek, Erik (2024). On the physical foundations of interstellar space travel. In: *Chronicles of Business Informatics Physics (CBIP)*. Volume 2, edition no. 1.0. ISBN: 9783759731159.

components of an electromagnetic field for the (general) case of the special theory of relativity correspond to all the very familiar Maxwell-Pointing equations.

Albert Einstein now had all the derived laws of nature in the most general form, which correspond to the gravitational field and the mass, because he logically used his coordinate reference system, with respect to which $\sqrt{-g}$ is equal to 1. He thus achieved a significant generalization of all equations and calculations, without neglecting the general covariance condition: he discovered his systems of equations by differentiating (specializing) his coordinate reference system using covariant general systems of equations.

In any case, a question of formal attention appears to be whether, by a concordant generalized determination of all energy components of the mass and the gravitational field, conservation axioms in the form of the equation system (56) and gravitational field equations in the class of equation systems (52) or (52a) are also valid without the differentiation of a coordinate reference system, so that on the left side the deviation (within its actual sense) is written, on the right side an addition of all energy components of the gravity and the mass. Albert Einstein had discovered that both conservation laws actually apply. However, he believed that publicizing his very extensive investigations into this subject would be of no benefit whatsoever because, contrary to his assumption, it would not lead to any new physical experience.

E. § 21. Newton's theory of gravity resembles a first approximation. {§ 21. Aspects of consideration with regard to the derivation of approximation-based valid laws of nature.}

What has often been said is that the special theory of relativity for the special case of the general (theory of relativity) is described by this, namely that all $g_{\mu\nu}$ have all constant quantities (4). According to the last statement, this means that gravitational influences are completely dispensed with. Albert Einstein determined his approximation, which is more closely related to what is happening, by investigating a

Kolek, Erik (2024). On the physical foundations of interstellar space travel. In: *Chronicles of Business Informatics Physics (CBIP)*. Volume 2, edition no. 1.0. ISBN: 9783759731159.

(physical) scenario in which all $g_{\mu\nu}$ only diverge close to (approximately 1) low values with regard to the quantities (4), whereby he dispensed with {infinite minimum} low values with the rank two and higher. (*The first consideration aspect for the approximation*).

In addition, it is theorized that within the observed (four-dimensional) time-space domains, all $g_{\mu\nu}$ within infinite space for the matching selection of all point coordinates approximate all quantities (4); this means Albert Einstein observed gravity fields that are only observable from the mass found within the non-infinite space.

In general, one would have to agree here and now, as the previous rejections {Albert Einstein} should lead to Newton's theory of gravitation. Instead, an additional approximate transformation of all initial system equations with the help of the second {further} aspect is necessary. Albert Einstein considered the mechanics of a point mass according to the system of equations (46). For the scenario of the special theory of relativity, all components

dx_1/ds, dx_2/ds, dx_3/ds (Einstein, 1916, p. 816)

have arbitrary quantities {, which are equal to a requirement $(dx_1/ds)^2 + (dx_2/ds)^2 + (dx_3/ds)^2 < 1$ (Einstein, manuscript, p. 54)}; that is, arbitrary velocities

$v = \sqrt{(dx_1^2/dx_4 + dx_2^2/dx_4 + dx_3^2/dx_4)}$ (Einstein, 1916, p. 816)

are possible, which are lower compared to the speed of light movement (v less than 1) in a vacuum. If a general conscious decision is made to limit oneself to the scenario that is actually only noticed through experience (seeing), so that v appears low with regard to the movement of light, then this means for all components

dx_1/ds, dx_2/ds, dx_3/ds (Einstein, 1916, p. 816) {each $\ll 1$ (Einstein, manuscript, p. 56)},

Kolek, Erik (2024). On the physical foundations of interstellar space travel. In: *Chronicles of Business Informatics Physics (CBIP)*. Volume 2, edition no. 1.0. ISBN: 9783759731159.

that these must be thought of as equal to low values, although $dx_4/ds = 1$ except for values after the second derivative. (*The second aspect for the approximation*).

Now Albert Einstein presupposed that according to the first aspect of consideration for the approximation all values $\Gamma^\tau_{\mu\nu}$ whose low values must be present at least as a first derivative. The view of (46) therefore trains, namely that according to this system of equations, in accordance with the second aspect of consideration for the approximation, only {the scenario v equals μ equals 4 comes into question} expressions must be included with regard to which v equals μ equals 4. For the containment of expressions of smallest {all first} derivative {referring to the two aspects of consideration}, the following systems of equations are generally set up initially as a replacement for (46)

$d^2x_\tau/dt^2 = \Gamma^\tau_{44}$ (Einstein, 1916, p. 817),

wherein *dt* equals *dx₄* equals *ds*, or with respect to the expressions which, according to the first aspect of consideration, are present for the approximation as the first derivative:

{$d^2x_\tau/dt^2 = [44; 1] + [44; 2] + [44; 3] - [44; 4]$ (Einstein, manuscript, p. 56)}

$d^2x_\tau/dt^2 = [44; \tau]$ ($\tau = 1, 2, 3$) (Einstein, 1916, p. 817)

$d^2x_4/dt^2 = - [44; \tau]$ (Einstein, 1916, p. 817).

If, in addition, it is generally assumed that a gravitational field exists as a minimally dynamic one, in that there is a general limitation to the scenario, so that a mass, which causes a gravitational field, progresses only leisurely (compared to the speed of light movement), then in general {a limitation to members} a neglect of orders according to the temporal and spatial point coordinates is conceivable with respect to the left side, thus generally resulting in

(67) $d^2x_\tau/dt^2 = - 1/2 \times \Omega g_{44}/\Omega x_\tau$ ($\tau = 1, 2, 3$) (Einstein, 1916, p. 817).

Kolek, Erik (2024). On the physical foundations of interstellar space travel. In: *Chronicles of Business Informatics Physics (CBIP)*. Volume 2, edition no. 1.0. ISBN: 9783759731159.

This corresponds to the system of equations for the mechanics of a mass point according to Newton's theory of gravitation, in which $g_{44}/2$ characterizes the gravitational power potential. The only strange thing about this result is that the element g_{44} of the basic model tensor causes this mass point mechanics separately within the first approximation.

For arbitrarily occurring velocities, which do not appear to be greater than the movement of light in a vacuum ($v < 1$), the following applies due to (67)

$v = \sqrt{(dx_\tau^2/dt)} = \sqrt{(-\ 1/2 \times \Omega g_{44}/\Omega x_\tau)}$ (Einstein, 1916, p. 817)

Albert Einstein now considered the gravitational field equations (53). When looking at it pictorially, it must be noted that the mass density ϱ within the actual sense approximately alone determines the energy model tensor for the "mass", i.e. by means of the second expression opposite the left-hand side of equation (58) [or (58a) or (58b)]. If the approximation requested by Albert Einstein is generally carried out, then every component is omitted except the component

$T_{44} = T = \varrho$ (Einstein, 1916, p. 817).

On the left-hand side at the second position in (53), a tiny expression arises in the second derivation; the former brings about the approximation demanded by Albert Einstein

$+\ \Omega/\Omega x_1[\mu, v;\ 1] + \Omega/\Omega x_2[\mu, v;\ 2] + \Omega/\Omega x_3[\mu, v;\ 3] - \Omega/\Omega x_4[\mu, v;\ 4]$ (Einstein, 1916, p. 818).

This results in v equals μ equals 4 if the expressions differentiated according to the time scale are omitted

$-\ 1/2 \times (\Omega^2 g_{44}/\Omega x_1^2 + \Omega^2 g_{44}/\Omega x_2^2 + \Omega^2 g_{44}/\Omega x_3^2 = -\ 1/2 \times \Delta g_{44}$ (Einstein, 1916, p. 818).

The last-placed of all equation systems (53) therefore yields

(68) $\Delta g_{44} = x\varrho$ (Einstein, 1916, p. 818).

Kolek, Erik (2024). On the physical foundations of interstellar space travel. In: *Chronicles of Business Informatics Physics (CBIP)*. Volume 2, edition no. 1.0. ISBN: 9783759731159.

Both systems of equations (67) and (68) together appear to be equivalent to Newton's theory of gravitation.

With regard to the gravitational power potential, according to (67) and (68) the element

(68a) $-(x/8\pi)\int(\varrho d\tau/r)$ (Einstein, 1916, p. 818)

although Newton's law of gravitation is based on a time scale selected by Albert Einstein

$-(K/c^2)\int(\varrho d\tau/r)$ (Einstein, 1916, p. 818)

respectively

$-(K/v^2)\int(\varrho d\tau/r) = -(K/(\sqrt{(-1/2 \times \Omega g_{44}/\Omega x_\tau)})^2)\int(\varrho d\tau/r)$ (Einstein, 1916, p. 818)

in which K (= G) characterizes a normally known gravitational invariant as a certain invariant 6.7×10^{-8}. By means of equation Albert Einstein obtained

(69) $x = 8\pi K/c^2 = 1{,}87 \times 10^{-27}$ (Einstein, 1916, p. 818)

respectively

(69a) $x = 8\pi K/v^2 = 8\pi K/(\sqrt{(-1/2 \times \Omega g_{44}/\Omega x_\tau)})^2$ (Einstein, 1916, p. 818)

Since this is the power potential of gravitation according to Newton, it can be set as a sequence K = G = F, analogous to the procedure described in section 20, generally resulting in the Newtonian-Maxwellian gravitational power potential

(69b) $x = 8\pi F/c^2 = 1{,}87 \times 10^{-27}$ (Einstein, 1916, p. 818)

respectively

(69c) $x = 8\pi F/v^2 = 8\pi F/(\sqrt{(-1/2 \times \Omega f_{44}/\Omega x_\tau)})^2$ (Einstein, 1916, p. 818).

Kolek, Erik (2024). On the physical foundations of interstellar space travel. In: *Chronicles of Business Informatics Physics (CBIP)*. Volume 2, edition no. 1.0. ISBN: 9783759731159.

§ 22. Behavior of the scales and time scales within the non-dynamic gravity field. Bending of light emissions. {Change of spectral light rays}. Perihelion orbit of all planetary motions.

So that Newton's gravitational mechanics can be determined within a first approximation, Albert Einstein only had to determine g_{44} based on the 10 components of the gravitational power potential $g_{\mu\nu}$, because only g_{44} is to be included within the first approximation (67) of the system of equations for the mechanics of a mass point within a gravitational field. In general, it is already recognized that further different components of all $g_{\mu\nu}$ with regard to the quantities noted within (4) have to differ within the first approximation, so that this inequality is given by the requirement {system equation} g equal to –1.

With regard to a material point that creates a gravitational field and is located within an original value of a coordinate reference system, a *non-anti-symmetrical* result concerning the radius generally arises within a first approximation

$$
\begin{array}{lll}
g_{\varrho\sigma} = -\,\delta_{\varrho\sigma} - \alpha(x_\varrho x_\sigma / r^3)\ (1 < \varrho\ \text{and}\ \sigma < 3) & \text{(Einstein, 1916, p. 819)} \\
(70) \quad \{ \quad g_{\varrho 4} = g_{4\varrho} = 0\ (1 < \varrho < 3) & \text{(Einstein, 1916, p. 819)} \\
\quad\quad\quad\ g_{44} = 1 - \alpha/r & \text{(Einstein, 1916, p. 819).}
\end{array}
$$

0 or 1 equals $\delta_{\varrho\sigma}$, depending on whether σ equals ϱ or $\sigma\varrho$ {σ not equal to ϱ}, the value r equals

$+ \sqrt{(x_1{}^2 + x_2{}^2 + x_3{}^2)}$ (Einstein, 1916, p. 819).

This appears due to (68a)

(70a) $\alpha = xM/8\pi$ (Einstein, 1916, p. 819),

as soon as a point of matter producing a gravitational field is characterized by M. Whether all gravitational field equations (not within matter) are confirmed within the first approximation by means of this result is easily verifiable.

Kolek, Erik (2024). On the physical foundations of interstellar space travel. In: *Chronicles of Business Informatics Physics (CBIP)*. Volume 2, edition no. 1.0. ISBN: 9783759731159.

Albert Einstein now considered an effect that space or its measured functions (i.e. the gravitational field) experience due to the mass field M. There is always a true relationship between all "local" (Section 4) metric spatial distances and time distances *ds* on the one hand and all coordinate point differences *dx$_v$* on the other hand

$$ds^2 = g_{\mu v}dx_{\mu}dx_v \text{ (Einstein, 1916, p. 819).}$$

With regard to a uniform rod dimension plotted "side by side" to the x-coordinate axis (i.e. a metric rod with, for example, the hundredth part of a meter as length), the following would have to be specified

$$ds^2 = -1; dx_2 = dx_3 = dx_4 = 0 \text{ (Einstein, 1916, p. 819),}$$

therefore

$$-1 = g_{11}dx_1{}^2 \text{ (Einstein, 1916, p. 819).}$$

(Depending on the angle of observation, the x-axis and the rod can also be seen superimposed on each other.) If (i.e.) the uniform rod dimension overlaps the x-coordinate axis at the same time, then the same equation of all systems (70) noted first corresponds to

$$g_{11} = - (1 + \alpha/r) \text{ (Einstein, 1916, p. 819).}$$

Based on these two relationships, a first approximation results in a sequence of exactly

$$(71) \; dx = 1 - \alpha/2r \text{ (Einstein, 1916, p. 819).}$$

A (actually) uniform rod dimension therefore looks truncated by its relationship with the coordinate symmetry system in accordance with the determined size due to the existence of a gravitational field as soon as this rod is placed in the same direction as the radius (at rest) on the radius.

Equivalently, the (time-spatial) rod coordinate distance within the tangential mass point direction is generally determined by generally assuming that

$ds^2 = -1$; $dx_1 = dx_3 = dx_4 = 0$; $x_1 = r$, $x_2 = x_3 = 0$ (Einstein, 1916, p. 820).

One consequence is

$(71a) - 1 = g_{22}dx_2^2 = - dx_2^2$ (Einstein, 1916, p. 820).

During the tangential reference body alignment, a {coordinate reference system} gravitational field of a material point therefore has no effect on the (temporal) spatial coordinate distance of a rod.

Accordingly, all Euclidean geometric properties become invalid within a gravitational field from the very first approximation, if in general one and the same rod body is to be understood detached with regard to its position and the associated direction corresponding to a realization of the same distance. Admittedly, (69) and (70a) teach that all determinable changes hardly appear large enough to be noticeable during a metric survey of smaller planetary surfaces such as that of the Earth.

Now, with regard to the time coordinate, a running speed of a clock hand that is mechanically identical to others and exists motionless within a non-dynamic gravitational field is considered. The following is valid with regard to the passage of time (or the time period, i.e. the time required for each complete rotation of the clock hand)

$ds = 1$; $dx_1 = dx_2 = dx_3 = 0$ (Einstein, 1916, p. 820).

Accordingly

$g_{44}dx_4^2 = 1$ (Einstein, 1916, p. 820);

$dx_4 = 1/\sqrt{[g_{44}]} = 1/\sqrt{[1 + (g_{44} - 1)]} = 1 - (g_{44} - 1)/2$ (Einstein, 1916, p. 820)

respectively

$(72)\ dx_4 = 1 + (x/8\pi)\int(\varrho d\tau/r)$ (Einstein, 1916, p. 820).

Kolek, Erik (2024). On the physical foundations of interstellar space travel. In: *Chronicles of Business Informatics Physics (CBIP)*. Volume 2, edition no. 1.0. ISBN: 9783759731159.

Accordingly, the running period of the hand clock (and of time in general) is faster as soon as it does not lie at rest near heavy matter {appears localized}. The (first) consequence of this is that all (bluish) ends of spectral lines with respect to the light emission lines reaching the Earth have to look the same with respect to the reddish beginnings of spectral lines starting from huge body surfaces of massive suns. [Equivalent to the investigated clock period, the second consequence with regard to the light coordinate is that red spectral ends must appear shorter as long as the light moves more slowly near a heavy mass, whereby, based on Albert Einstein, the law of the constant speed of light no longer appears to be tenable, at least on the basis of (72); ponderable point masses therefore have a slowing effect on the light period, so this is also to be considered equivalent to the time period (72). Consequently, test reference bodies accelerated on geodesic spectral lines must always stay away from ponderable matter.] Pro of the existence of this effective interaction between light and mass confirm line spectral sightings of fixed stars of classified class, in agreement with Freundlich, E.. The final verification of the first sequence is no longer open today.

Albert Einstein also considered the movement of all light emissions within the non-dynamic gravitational field. In accordance with the special theory of relativity, the speed of movement of light can be calculated using the system of equations

$$- dx_1{}^2 - dx_2{}^2 - dx_3{}^2 + dx_4{}^2 = 0 \text{ (Einstein, 1916, p. 821)}$$

respectively

$$+ dx_1{}^2 + dx_2{}^2 + dx_3{}^2 - dx_4{}^2 = 0 \text{ (Einstein, 1916, p. 821)},$$

according to the general theory of relativity with the system of equations

$$(73) \quad ds^2 = g_{\mu\nu}dx_\mu dx_\nu = 0 \text{ (Einstein, 1916, p. 821)}$$

respectively

$$(73a) \quad ds = \sqrt{(g_{\mu\nu}dx_\mu dx_\nu)} = 0 \text{ (Einstein, 1916, p. 821)}.$$

Kolek, Erik (2024). On the physical foundations of interstellar space travel. In: *Chronicles of Business Informatics Physics (CBIP)*. Volume 2, edition no. 1.0. ISBN: 9783759731159.

If there is an orientation, i.e. the relation dx_1 to dx_2 to dx_3, then the system of equations (73) has the following values

dx_1/dx_4, dx_2/dx_4, dx_3/dx_4 (Einstein, 1916, p. 821)

and the resulting speed of light $x_1 = v$

$$v = \sqrt{[(dx_1/dx_4)^2 + (dx_2/dx_4)^2 + (dx_3/dx_4)^2]} = \sqrt{[(dx_\mu/dx_\nu)^2]} \text{(Einstein, 1916, p. 821)},$$

respectively

$$v^2 = (dx_1/dx_4)^2 + (dx_2/dx_4)^2 + (dx_3/dx_4)^2 = (dx_\mu/dx_\nu)^2 \text{ (Einstein, 1916, p. 821)},$$

are determined in accordance with Euclid with respect to a space-time geometry. In general, it is easy to understand that all light emission lines must look curved with respect to the coordinate reference system if all $g_{\mu\nu}$ appear variable. If x_2 represents the vertical orientation with respect to the light movement, then Huygens' principle shows that the light emission line [observed within a two-dimensional surface (x_1, x_2)] has a bend $- \Omega x_1/dx_2$.

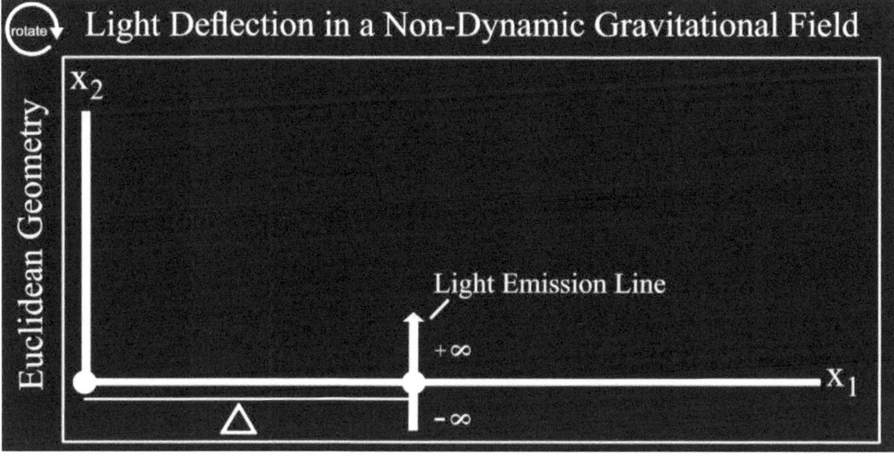

Figure 1. Light deflection in a non-dynamic gravitational field corresponding to a two-dimensional space-time geometry based on Euclid (Einstein, 1916, p. 821).

Kolek, Erik (2024). On the physical foundations of interstellar space travel. In: *Chronicles of Business Informatics Physics (CBIP)*. Volume 2, edition no. 1.0. ISBN: 9783759731159.

Albert Einstein considered a bend experienced by a light emission line moving along the (heavy) matter M within the distance Δ. If a coordinate reference system is generally selected in accordance with the previous figure, then a complete curvature B of the light emission line (calculated non-negatively if the curvature is not curved (concave) in front of and towards its point of origin) appears within a sufficient approximation described by

$$B = \int_{-\infty}^{+\infty} \frac{\Omega x_1}{\Omega x_1} dx_2 = \int_{-\infty}^{+\infty} dx_2 \text{ (Einstein, 1916, p. 821)}$$

meanwhile (70) and (73) have the following result

$x_1 = \sqrt{(-g_{44}/g_{22})} = 1 + 1/2 \times \alpha/r[1 + (x_2/r)^2]$ (Einstein, 1916, p. 822).

The following is obtained through dissolution

(74) $B = 2\alpha/\Delta = xM/4\pi\Delta$ (Einstein, 1916, p. 822),

(74a) $M = 2\alpha/\Delta = 4\pi\Delta B/x$ bzw. $x = 2\alpha/\Delta = 4\pi\Delta B/M$ (Einstein, 1916, p. 822) and

(74b) $\pi = 2\alpha/\Delta = xM/4\Delta B$ (Einstein, 1916, p. 822).

The light emission line passing alongside a star therefore experiences a curvature of (approximately) 1.7" (angular seconds), while the line passing the non-light-emitting gas-dust-atmosphere rocky core planet Jupiter experiences a corresponding curvature of approximately 0.02" (arc seconds).

If a gravitational field and a course motion of a point mass with relatively infinitely small inertia are calculated more precisely with a factor, then Albert Einstein obtained the change of subsequent class for all planets compared with all Kepler-Newton motion theorems. A course ellipse {major axis} of an inertial reference body, for example of a planet or, on the other hand, of a light material point, undergoes rotation in the same direction of movement of the course {per orbit} at low speed given by the absolute value (connected with 74b)

Kolek, Erik (2024). On the physical foundations of interstellar space travel. In: *Chronicles of Business Informatics Physics (CBIP)*. Volume 2, edition no. 1.0. ISBN: 9783759731159.

(75) $\varepsilon = 24\pi^3[a^2/T^2c^2(1 - e^2)]$ (Einstein, 1916, p. 822)

(75a) $\varepsilon = 24(xM/4\Delta B)^3[a^2/T^2v^2(1 - e^2)]$ (Einstein, 1916, p. 822)

(75b) $\varepsilon = 24(xM/4\Delta B)^3[a^2/T^2(dx_\mu/dx_\nu)^2(1 - e^2)]$ (Einstein, 1916, p. 822)

per orbit. Within this system of equations, a stands for a main axis, c or v for the relative speed of light within the finitely constant or infinitely variable unit of measurement, e for the elliptical eccentricity, T for the orbiting time given in seconds. With regard to a calculation, Albert Einstein referred to two original papers (Einstein, 1916, p. 822): "A. Einstein, Sitzungsber. d. Preuß. Akad. d. Wiss. 47. p. 831. 1915. – K. Schwarzschild, Sitzungsber. d. Preuß. Akad. d. Wiss. 7. p. 189. 1916."

The result of this calculation with regard to the planetary body Mercury is that the rotation of its ellipse is (approximately) 43" (angular seconds) every hundred years, exactly in accordance with a finding by astronomical physicists (here Leverrier); the astrophysicists discovered, for example, the residual amount for the perihelion mechanics of Mercury with the quantified value (43"), which is by no means comprehensible with the influence of other planetary bodies.

§ 23. Gravitational-magnetic equations of the vacuum field according to Albert Einstein. Behavior of the scales and time scales within the non-static gravitational field.

If the electromagnetic equations of the vacuum field according to Maxwell are now equated with the gravitational-magnetic equations of the vacuum field according to Albert Einstein, the following is determined according to the *(anti-)general theory of relativity*

(66b) $T_\sigma^\nu = - G_{\sigma\alpha}G^{\nu\alpha} + 1/4 \times \delta_\sigma^\nu G_{\alpha\beta}G^{\alpha\beta}$ (Einstein, 1916, p. 815)

respectively *anti-general*

(66bb) $- T_\sigma^\nu = G_{\sigma\alpha}G^{\nu\alpha} - 1/4 \times \delta_\sigma^\nu G_{\alpha\beta}G^{\alpha\beta}$ (Einstein, 1916, p. 815).

Kolek, Erik (2024). On the physical foundations of interstellar space travel. In: *Chronicles of Business Informatics Physics (CBIP)*. Volume 2, edition no. 1.0. ISBN: 9783759731159.

Substituting (66) for electromagnetic gravitational fields or gravitating electromagnetic fields results in

(66c) $x_\sigma = \Omega(- G_{\sigma\alpha}G^{v\alpha} + 1/4 \times \delta_\sigma{}^v G_{\alpha\beta}G^{\alpha\beta})/\Omega x_v - 1/2 \times g^{\tau\mu}(\Omega g_{\mu v}/\Omega x_\sigma)(- G_{\tau\alpha}G^{v\alpha} + 1/4 \times \delta_\tau{}^v G_{\alpha\beta}G^{\alpha\beta})$ (Einstein, 1916, p. 815)

respectively *anti-general*

(66cc) $- x_\sigma = - \Omega(G_{\sigma\alpha}G^{v\alpha} - 1/4 \times \delta_\sigma{}^v G_{\alpha\beta}G^{\alpha\beta})/\Omega x_v + 1/2 \times g^{\tau\mu}(\Omega g_{\mu v}/\Omega x_\sigma)(G_{\tau\alpha}G^{v\alpha} - 1/4 \times \delta_\tau{}^v G_{\alpha\beta}G^{\alpha\beta})$ (Einstein, 1916, p. 815)

Looking back to the beginning of this section 20, each F of gravitating electromagnetic fields can therefore be replaced by a G of electromagnetic gravitational fields, whereby the power potential of the (also electric) gravitational magnetism is described as a consequence due to $F_{\varrho\sigma} = G_{\varrho\sigma}$. For example, all energy components of the gravitational magnetic field can be determined, which are particularly relevant for smaller gravitational magnetic fields that are dependent on larger gravitational magnetic fields over time.

(65aa) $x_\sigma = G_{\sigma\mu}J^\mu$ (Einstein, 1916, p. 814).

For the physical *anti-general case* of dynamic (unnatural) gravitational fields, the variable j can be used to metrically characterize the instantaneous power potential of the electric gravitational magnetic field, so that a certain mass gravity (equal to mass inertia) is always given within the four-dimensional space-time domain of uniform curvature for all practically rigid bodies located therein. The following system equations are coincident with electromagnetically moving gravitational fields due to the initial now dynamized requirement {reference system} jg = −1; all previous statements regarding the equations remain unchanged.

$$(70a) \quad \begin{cases} j_{\varrho\sigma}g_{\varrho\sigma} = -\delta_{\varrho\sigma} - \alpha(x_\varrho x_\sigma/r^3) \ (1 < \varrho \text{ and } \sigma < 3) & \text{(Einstein, 1916, p. 819)} \\[6pt] j_{\varrho 4}g_{\varrho 4} = j_{4\varrho}g_{4\varrho} = 0 \ (1 < \varrho < 3) & \text{(Einstein, 1916, p. 819)} \\[6pt] j_{44}g_{44} = 1 - \alpha/r & \text{(Einstein, 1916, p. 819).} \end{cases}$$

$(71aa) - 1 = j_{22}g_{22}dx_2^2 = -dx_2^2$ (Einstein, 1916, p. 820).

$(72a) \ dx_4 = 1/\sqrt{[j_{44}g_{44}]} = 1/\sqrt{[1 + (j_{44}g_{44} - 1)]} = 1 - (j_{44}g_{44} - 1)/2$ (Einstein, 1916, p. 820)

$(73a) \ ds^2 = j_{\mu\nu}g_{\mu\nu}dx_\mu dx_\nu = 0$ (Einstein, 1916, p. 821)

The latter equations show that a static energy field such as the matter field and gravitational field could only appear to exist within the entire space-time continuum, which means that all matter fields and gravitational fields would always have to exist dynamically and constantly influence each other. This is also justified by the gravitational-magnetic equations of the vacuum field according to Albert Einstein, within which the laws of electrostatics and electrodynamics are also taken into account. It follows that every mass with a negative electric charge must be attracted by gravity and that every mass with a positive electric charge must be repelled by *anti-gravity*.

This is also due to the spin (angular momentum) of the point mass, which should lead to an electrically positive charge of the same mass. At the two practically rigid poles of this rotating and therefore moving point mass, electrically positively charged bodies (protons) are therefore radiated or emitted. In the vacuum of the space-time continuum, this positive proton radiation undergoes a discharge and reversal of direction in the physical sense, whereby negatively charged bodies (electrons) fall back onto the point mass or are attracted by it, which closes the cycle of the energy field consisting of matter field and gravitational field in an *anti-general* way, since negative electron radiation is constantly present due to gravitational-magnetic general laws of nature and is absorbed by the point mass.

Kolek, Erik (2024). On the physical foundations of interstellar space travel. In: *Chronicles of Business Informatics Physics (CBIP)*. Volume 2, edition no. 1.0. ISBN: 9783759731159.

Protons and electrons can therefore also be understood as electromagnetized gravitational particles with different positive or negative spin charges, analogous to the point mass nucleus, with the result that lighter electrons could possibly only emit photons during their discharge and reversal of direction, while heavier protons would only have to attract photons. The probably only apparently existing static magnetic field of a point mass therefore resembles a physical interaction of emitting positive proton radiation and absorbing negative electron radiation and thus the dynamic gravitational field of the same point mass field. An *anti-gravitational field* of a body can therefore arise as soon as the negative electron radiation experiences a positive sustained energy pulse, which is also conceivable if the two poles are shifted. According to the *anti-general principle*, such a pole shift is theoretically possible as soon as the body's angular momentum decreases, which also means that the temperature and velocity of the mass must decrease.

If the point mass is considered as a whole, then it should consist of a hot core and at least one surrounding cooler shell, between which there could be *anti-symmetrical* vacuum gaps, which could also cause further shells as well as differences in rotational temperature and rotational speed. The faster spin of the nucleus should therefore cause a slower spin of the outer shell and could also positively charge it electrically, but the greater the distance to the nucleus, the more the negative electron charge should be detectable and thus *anti-general gravity* should be experienced. In the opposite physical sense, however, this would also mean that in the vicinity of the nucleus (also due to the vacuum that is probably located around it) there would have to be a hovering point free of matter fields and gravitational fields ($g_{ik} = 0$ and respectively $\sqrt{-g} = 0$) in a certain four-dimensional space-time range in the sense of an extended event horizon surrounding the nucleus.

Simply put: *Gravity would have to decrease towards the core and gravity would have to increase (again) away from the core, but again the greatest gravity would therefore have to be present in the core, since the greatest punctual mass temperature gravity*

Kolek, Erik (2024). On the physical foundations of interstellar space travel. In: *Chronicles of Business Informatics Physics (CBIP)*. Volume 2, edition no. 1.0. ISBN: 9783759731159.

(equal to mass temperature inertia) in the physical sense of energy would have to be found here, around this ponderable core, in turn, an extended spatiotemporal event horizon would have to exist in the vicinity of the matter field, which physically would have to allow no time but also no movement of positively charged (dark) and negatively charged (light) matter in space, but could only generate positive (dark) and negative (light) energy in an infinite number of parallel dimensions (73a).

Thus, within our space-time continuum, only light-absorbing dark hot proton nucleus masses and light-giving bright cool electron nucleus masses could exist as dark or bright forms of energy, whereby $E = TMv^2 = G$ would have to apply within the described event horizon corresponding to the positive or negative gravitational energy charge, which also consequently justifies a substitution of the sign. Protons $(+\,G)$ should therefore attract photons and electrons $(-\,G)$ should therefore repel photons, whereby a single high-energy proton should be found in the nucleus of every low-energy electron as a consequence down to the infinitely small.

Whether this assumption proves to be true up to the infinitely large remains to be seen today due to a lack of practical space-time research experience and because this is a new task for the newly established macro-quantum physics (quantum physics II) analogous to the familiar micro-quantum physics (quantum physics I). In this just established overall model, today's astrophysics is placed between these two quantum physics maps as a link between these two dimensionally different quantum theories to complete physics or the model of everything.

Overall, the basis of the *anti-general theory of relativity* is therefore a *dynamic gravitational field theory* in which electrostatic and electrodynamic mass effects are transferred coincidentally to gravitational-static and gravitational-dynamic mass effects, making it possible to experience the gravitational field by equating energy and mass with regard to the temperature shift of the mass, which is similar to the spectral red shift of light discovered by Albert Einstein. Artificial generation of a dynamic

Kolek, Erik (2024). On the physical foundations of interstellar space travel. In: *Chronicles of Business Informatics Physics (CBIP)*. Volume 2, edition no. 1.0. ISBN: 9783759731159.

gravitational field within a dx_n-determined space-time body should now be possible using quantum technology on the basis of the assumptions described.

A final side note on the false numerical theory of relativity. Today's mathematics is generally an (imaginary) formal science and today's physics is an (observable) real science, which means that both types of science are not completely coincidentally applicable to each other, at least as long as mathematics does not represent a real science, because the (previously formal) mathematics only becomes real when it is physically transferred to the observable optical dynamics of light. To put it even more simply, this only means that numerical symbols conceived by humans cannot be applied directly to the space-time continuum or to the movements of bodies without a harmonizing adaptation of this mathematical theory by means of the optical dynamics of light.

As a first approximation, I imagine an arbitrarily determined scale, for example a meter of a light beam, then I already know from Albert Einstein's special and general theory of relativity that this light meter is not straight but curved and measured at its ends must not be uniform but of different lengths within the gravitational field of a body such as the earth. This is where arithmetic (number theory) already fails; the greater the deviation, the stronger the gravitational fields. For example, a light meter on Earth is curved heterogeneously compared to a light meter on Mars and placed on top of each other within a common coordinate system results in a different numerical value on the x-axis. The arithmetic therefore does not resemble an axiom according to the special and general theory of relativity. According to the *anti-general theory of relativity*, gravitational fields are always dynamic, which also does not change the physical fact that the arithmetic appears incorrect according to the optical dynamics of the light field.

However, another non-physical but purely mathematical proof that formal mathematics in the form of number theory known as arithmetic cannot work in real physics is also easily comprehensible from Pierre de Fermat's last theorem. Fermat

Kolek, Erik (2024). On the physical foundations of interstellar space travel. In: *Chronicles of Business Informatics Physics (CBIP)*. Volume 2, edition no. 1.0. ISBN: 9783759731159.

(just like me) did not need another theorem but only a last theorem to disprove arithmetic (number theory). Fermat's theorem states that $x^n + y^n$ is always unequal to z^n if $n = 3$ to infinity. Every schoolchild knows this theorem as the Pythagorean theorem, but it is only valid up to $n = 2$. This last theorem of Fermat can only be solved analytically and not numerically in the following way.

The characters x, y, z, n are different letters (imagined by humans) that are supposed to represent different numerical values (imagined by humans). Since I have to consider all numerical values and am not allowed to change the condition of inequality for this complete proof (i.e. not just provide a partial proof, for example like Euler for $n = 3$), I can now express that "infinity" as a sequence of letters is an arbitrarily determined number. If I insert this new number of letters into Fermat's last theorem, I obtain the following in general notation

$$\text{infinity}^{\text{infinity}} + \text{infinity}^{\text{infinity}} \neq \text{infinity}^{\text{infinity}}$$

respectively

$$2 \times \text{infinity}^{\text{infinity}} \neq \text{infinity}^{\text{infinity}}$$

or for all conceivable numbers quod erat demonstrandum

$2 \neq 1$ (The true proof of Fermat's last theorem, since the proof itself is an axiom.)

Every theory of relativity can therefore only ever be understood by mathematical physics analytically via algebra as a basis of geometry and never numerically via arithmetic, since this always represents a falsity, as has just been proven several times.

(Completed 07. June 2024.)

References

Einstein, A. (1916). Die Grundlage der allgemeinen Relativitätstheorie. *Annalen der Physik 354(7)*, pp. 769–822.

Kolek, Erik (2024). On the physical foundations of interstellar space travel. In: *Chronicles of Business Informatics Physics (CBIP)*. Volume 2, edition no. 1.0. ISBN: 9783759731159.

{Einstein, A. (manuscript). Die Grundlage der allgemeinen Relativitätstheorie. In: The General Theory of Relativity. SP Books, limited edition number 411/1000, pp. 11–62.}

Table of contents

In this research article, an anti-general theory of gravity is developed. It is referred to as the anti-general theory of relativity and is based on Albert Einstein's general theory of relativity. It contains a dynamic gravitational field theory in which several fields can influence each other. Spatial movements can thus be better described.

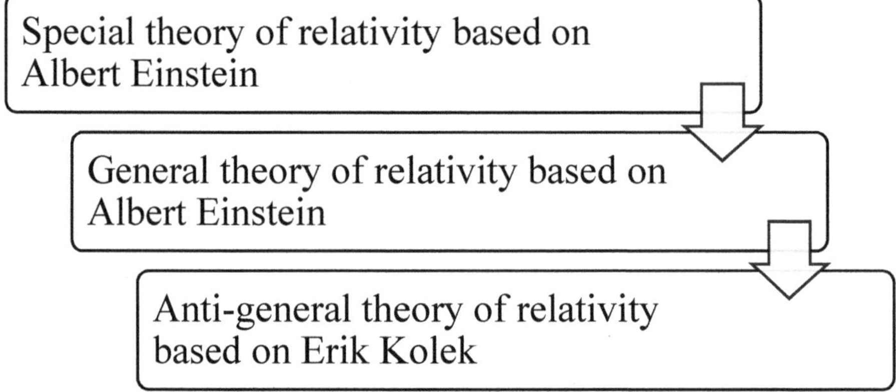

Figure 2. Table of contents illustrated by the significance of the anti-general theory of relativity.

Kolek, Erik (2024). On the physical foundations of interstellar space travel. In: *Chronicles of Business Informatics Physics (CBIP)*. Volume 2, edition no. 1.0. ISBN: 9783759731159.

Third section: A view into the darkness of the space-time continuum

Scientific citation:

Kolek, Erik (2024). Black holes have no gravity at their event horizon in our expanding universe – My first contribution to the Einstein-Hawking-Theory of everything. In: *On the physical foundations of interstellar space travel.* Chronicles of Business Informatics Physics (CBIP). Volume 2, edition no. 1.0.

Erik Kolek (2024)

Black holes have no gravity at their event horizon in our expanding universe – My first contribution to the Einstein-Hawking-Theory of everything

Summary

This book contribution is about the purely theoretical possibility that black holes could have no gravity at their event horizon in our expanding universe. This is my first contribution to the theory of everything, which I call the Einstein-Hawking-Theory, and with which I would like to give everyone a visual insight into the darkness of the space-time continuum. So are we inside a single giant black hole that represents our universe as a whole, or is this idea of the universe even conceivable? If such an infinitely manifold intensity of curvature should also be conceivable inwards, in which individually contravariantly connected space-time bubbles could open up clockwise periodically: What other types of gravitational fields exist within our possibly Black Universe and how can these findings be used for further developments of quantum technologies, for example to be constructively capable of equipping sovereign reference body classes (spaceship classes) with an even more modern energy momentum drive in terms of time-space? Supergravity could provide the answer if a moving space-in-space model tensor according to Riemann $\Gamma_{\mu\nu}{}^{\tau}$ is

Kolek, Erik (2024). On the physical foundations of interstellar space travel. In: *Chronicles of Business Informatics Physics (CBIP)*. Volume 2, edition no. 1.0. ISBN: 9783759731159.

followed, but not without balancing the mass inertia in this electrodynamic field bubble sg_{ik} in a heavy field generating g_{ik}.

Black holes have no gravity at their event horizon in our expanding universe – My first contribution to the Einstein-Hawking-Theory of everything

Black holes exist in our expanding universe, as observers from Earth already know, because these observers from Earth have taken a photo of such a black hole [this photo can be seen on the following website: https://eventhorizontelescope.org/press-release-april-10-2019-astronomers-capture-first-image-black-hole (last visited on 07.06.2024)]. This photo shows an accretion disk of a black hole, this disk resembles a boundary consisting of heat of light. These observers, mostly astrophysicists and cosmologists, work within the standard model of physics, but they also try to work outside the standard model of physics, although they are not yet ready to understand everything in the universe, such as nuclear fusion within stars. Within this standard model of physics, gravity is described as a field by Albert Einstein (1916). For these observers, gravity generally exists inside, at the event horizon and outside a black hole.

But what did the observers see in physical reality? Did the observers see a black hole vortex or a black hole tornado? Therefore, if I think multidimensionally of a curved singularity like a black hole, it is perhaps possible that two kinds of such curved singularities could exist. The important question then for all observers from Earth is on which side is our expanding universe localized? Are we perhaps on the side where matter is being pulled into a black hole or possibly on the side where matter is flowing out of that black hole? Based on this photo of a black hole, I cannot answer this question as an observer of the heat of light that I can see shining from Earth. And of course the physical fact remains in both general cases that a black hole in the

Kolek, Erik (2024). On the physical foundations of interstellar space travel. In: *Chronicles of Business Informatics Physics (CBIP)*. Volume 2, edition no. 1.0. ISBN: 9783759731159.

geometric form of a vortex or tornado can easily destroy a star, collide with other black holes and explode.

Side note. The latter also suggests that perhaps even electrodynamic weather phenomena such as ion storms could exist within our expanding universe, which is important to know for the reference body navigation of period research practice.

As I said before, observers cannot be sure if they see a black hole that emits and/or absorbs matter. Referring to Albert Einstein (1905, 1916), such singularities can only be curved and therefore never appear straight (or flat). Therefore, I completely exclude non-curved singularities and furthermore, curved black holes (black singularities) are the focus of consideration (Figure 2). Curved singularities can exist moving in two ways if our expanding universe is supersymmetric, relativistic: the first way on the left is described by the standard model of physics and the second way on the right represents a supersymmetric, relativistic mirror model dimension. And once again, both arrows point in the same direction the path of the heat of light as all of us on Earth have seen in the photo of a black hole. Of these two curved singularities, only one can exist in the physical reality of our expanding universe.

The model shown in Figure 1 is, as I assume for our supersymmetric, relativistic universe, coincident with my supersymmetric, relativistic modelling and visualization approach and it demonstrates possible model results for the development of precise mathematical physics. Because human multidimensional thinking is visually multimodal, i.e. complex, I have developed and used an approach to make visually multimodal, i.e. complex, physical facts easy to understand. With this approach, you do not need in-depth mathematical knowledge to be able to understand designed model components of the universe. Therefore, quantum astrophysics as a whole should become easier to understand. In addition, I use theorems and equations to describe model-based representations of our expanding universe. A collection of model visualizations is necessary to make it possible to experience the entire universe down-to-earth in a multidimensional way.

Kolek, Erik (2024). On the physical foundations of interstellar space travel. In: *Chronicles of Business Informatics Physics (CBIP)*. Volume 2, edition no. 1.0. ISBN: 9783759731159.

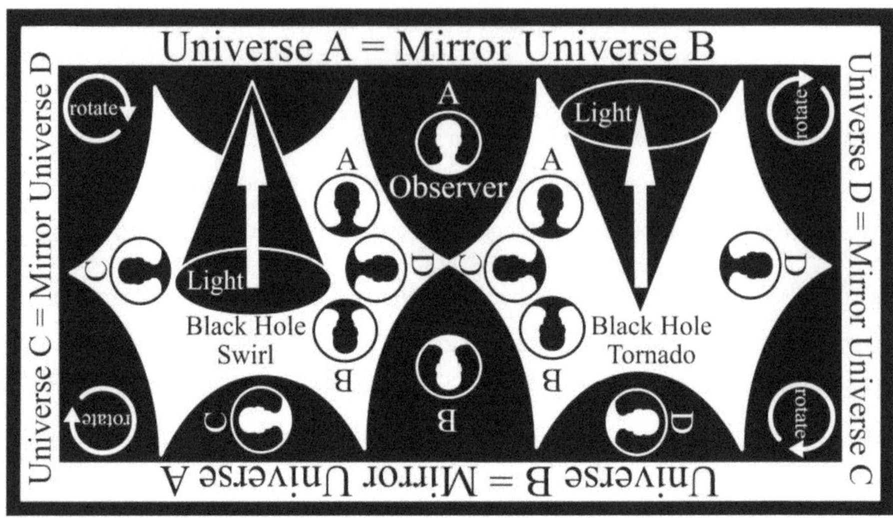

Figure 1. Possibility of two curved singularities (which one exists in our expanding universe?).

For each type of black hole, the quantum spectrum of emitting and/or absorbing matter is thermal and given for the emission as a generally accepted (in quantum physics) blackbody radiation temperature – the Hawking temperature is given in natural units and $T_H = k / 2\pi$ (Hawking, 1974; Hawking, 1975). In the International System of Units (abbreviated SI units) [see: https://www.bipm.org/en/measurement-units/ (last visited on 07.06.2024)] this equation converts to $T_H = hk / 2\pi c k_B$ (Hawking, 1974; Hawking, 1975). The Schwarzschild gravity for the surface of a black hole is represented in SI units by the equation $k = c^4 / 4GM$ (Schwarzschild, 1916). With respect to the mass M of a black hole, the Hawking temperature is then inversely proportional in equation (6). The Hawking temperature therefore includes the reduced Planck constant (h), the light (heat) constant (c), gravitational constant (G), (black hole) mass (M) and the Boltzmann constant (k_B) (Hawking, 1974; Hawking, 1975).

(6) $T_{H1} = hc^3 / 8\pi GMk_B$ (Mass reduction leads to temperature rise) (Hawking, 1974; Hawking, 1975)

Kolek, Erik (2024). On the physical foundations of interstellar space travel. In: *Chronicles of Business Informatics Physics (CBIP)*. Volume 2, edition no. 1.0. ISBN: 9783759731159.

Consequently, the Hawking temperature (Hawking, 1974; Hawking, 1975) appears to be consistent with the model assumption: if a black hole emits the energy L in the form of mass and radiation, its mass decreases by L/V^2 (Einstein, 1905). For this special case, the Hawking temperature increases (Hawking, 1974; Hawking, 1975), but it should also increase if a black hole absorbs the energy L in the form of mass and radiation, its mass increases by L/V^2 (Einstein, 1905). Therefore, I obtain a second Hawking temperature (7) (Hawking, 1974; Hawking, 1975)

(7) $T_{H2} = 8\pi GMk_B / hc^3$ (mass increase leads to temperature rise) (Hawking, 1974; Hawking, 1975)

With respect to a black hole that absorbs and emits mass and radiation (Einstein, 1905), the completed Hawking temperature (Hawking, 1974; Hawking, 1975) is given in equation (8). All links are now based on an advanced theory of the heat of light supersymmetrically and relativistically able to explain temperature changes at the event horizon of a black hole via its mass adjustment (or mass substitution).

(8) $T_{H3} = hc^3 / 8\pi GMk_B + 8\pi GMk_B / hc^3$ (mass change leads to temperature change) (Hawking, 1974; Hawking, 1975)

Equation (8) can be used to derive new, more modern model assumptions relevant to our expanding universe. Black holes cannot have gravity G at their event horizon in our expanding universe due to hypotheses (a), (b), (c) and (d).

(a) Every black hole grows or expands in our universe, but they still do not draw matter into themselves. So what does a black hole represent in our expanding universe, a black hole vortex or a black hole tornado?

(b) Every black hole has some kind of gravity inside and no kind of gravity outside.

(c) As a further cross-verification for (a) and (b) leading to (d) would have to apply: If the Ferent theory of gravity (2016) – whose basis is still Newton's theory of gravity (1687), because Ferent (2016) believes that Albert Einstein's special and general

Kolek, Erik (2024). On the physical foundations of interstellar space travel. In: *Chronicles of Business Informatics Physics (CBIP)*. Volume 2, edition no. 1.0. ISBN: 9783759731159.

theory of relativity (1905, 1916) would be limited with respect to the speed of light V – should be true, the Ferent theory of gravity (2016) should also apply to black holes and their event horizons.

(d) With respect to a black hole and its event horizon, gravity G can be discarded and therefore removed from the Hawking temperature equations (6), (7) and (8) (Hawking, 1974; Hawking, 1975). This should be possible on the basis of an advanced theory of the heat of light, which appears as soon as the special theory of relativity of Albert Einstein (1905) is thought further as follows and on the basis of (a) and (b).

Let us now imagine a dark universe consisting only of black holes and matter (Figure 2). What would the observer see? He or she must be able to see within a black hole any number of other black holes and a black event horizon within the manifold hollow ellipsoid, where he or she is localized on a planet and looks at its event horizon from within its black hole, but which appears only black and not glowing black-red due to the distance of the spectral lines. So that this black hole behaves like our universe in that the boundaries seen from within (apparently) expand, so each black hole could also form its own universe in a hollow curvature. Consequently, this black hole grows as seen by another observer from outside this black hole or dark universe. Here again, our black hole lies inside another black hole and so on into the infinity of the manifold parallelism of the dark universe. From above, in this macro view of the physics of the universe (Quantum Astrophysics II), the observer sees in this special case only infinitely smaller black holes $dx_1 + dx_2 + dx_3 + dx_4$ compared to his dark universe. In this physical model example, four black holes and four event horizons are selected because this number of quantities should be sufficient to understand the infinity of a universe consisting only of curved singularities. How does gravity change in the geometric form of a field g_{ik} as we have all learned from Albert Einstein and Stephen Hawking? In the space x, y, z, time t stops and becomes absolute and zero at each event horizon of the (four) black holes, but now also the gravitational field $g_{ik} = dt = dx_4 = 0$ at each event horizon of all black holes. This model comparison between our

Kolek, Erik (2024). On the physical foundations of interstellar space travel. In: *Chronicles of Business Informatics Physics (CBIP)*. Volume 2, edition no. 1.0. ISBN: 9783759731159.

dark universe seen as a single giant black hole should help to understand what is meant by propositions when talking about existence within an expanding black hole, in which a special gravitational field should also exist just as in our expanding universe.

So what is a supergravity field? A supergravity field (sg_{ik}) is described by the sum of the gravitational fields g_{ik} of all mass fields and radiation energies that can exist within a dark universe or black hole (contravariant). In this context, a black hole naturally represents a universe. According to nature, supergravity also includes the inner effect relationship of the same shrinking or expanding dark universe or black hole, because this special gravitational field also shrinks or expands uniformly in coincidence with it. Therefore, supergravity is connected with all moments of simultaneity that can exist manifold in parallel within the space-time continuum and for the model simplification of our clock period, $sg_{ik} = dt = dx_4 = 1$. In this universe, a supergravity field exists between all black holes. If $sg_1 = sg_2 = sg_3 = sg_4$ in this universe, then all black holes do not grow or shrink. So how do black holes grow or shrink? Black holes grow or shrink if $sg1_{ik} > sg2_{ik} > sg3_{ik} > sg4_{ik}$. In this general case, an observer sees a black hole tornado consisting of matter (this is also comparable to the aether theory). Black holes shrink or become smaller if $sg1_{ik} < sg2_{ik} < sg3_{ik} < sg4_{ik}$. In this general case, an observer sees a black hole vortex consisting of matter (this is also analogous to the aether theory).

In our expanding universe, observers see growing black holes and the boundary of our universe appears to be expanding outwards like a hollow ellipsoid, which is why a supergravitational field $sg1_{ik} > sg2_{ik}$ can generally be assumed for our dark universe for the black holes inside it. The observers only see a kind of curved singularity – the black hole tornado that whirls matter around. A black hole tornado can therefore also destroy a star simply thanks to its physics, collide with other matter tornadoes and explode. Such a matter tornado looks exactly the same from above in two dimensions as in the photo of the black hole that has already been taken by observers on Earth.

Kolek, Erik (2024). On the physical foundations of interstellar space travel. In: *Chronicles of Business Informatics Physics (CBIP)*. Volume 2, edition no. 1.0. ISBN: 9783759731159.

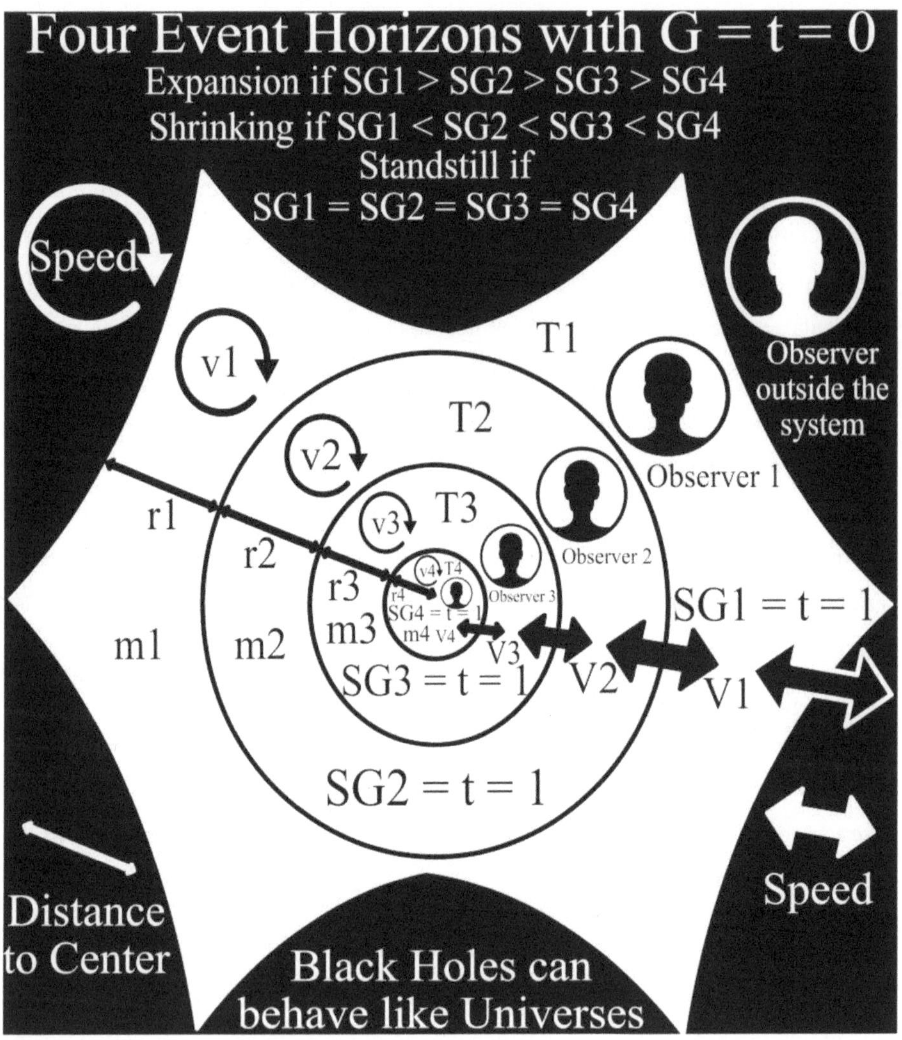

Figure 2. Supergravity fields derived from infinite internal black holes and their event horizons assumed as dark universe boundaries.

To provide a well-founded theory, supergravity fields can also be explained by a theory of light heat (Figure 3) based on the principle of Albert Einstein (1905; 1916), because the assumed speed of light c does not change during the emission and

Kolek, Erik (2024). On the physical foundations of interstellar space travel. In: *Chronicles of Business Informatics Physics (CBIP)*. Volume 2, edition no. 1.0. ISBN: 9783759731159.

absorption of light heat. Therefore, I use equation (9) as a basic postulate for the theory extension that leads to a unified field theory of gravitation.

(9) $K_0 - K_1 = L\,[(1\,/\,\sqrt{(1 - (v/V)^2)}) - 1]$ (Einstein, 1905)

Subsequently, I have to think of two coordinate systems in relative motion and of the speed of light radiation heat expressed as the constant c, because I can now assume fundamentally thanks to Einstein (1916) that everything in the universe is quantum matter (including electromagnetic fields) except the special gravitational field existing as a single finite infinite supergravitational field, which in truth always appears to be present in our apparently expanding universe as well as in relativistically based black holes according to general laws of nature. As a consequence, I obtain a new Einstein field equation that describes a unified gravitational field theory derived from the physical principle using the speed of light heat c (10), which explains two types of light radiation heat – bright and dark light represented by colliding bright and dark photons. From the perspective of quantum physics, these can also be bright white quantum stars and dark black quantum holes. However, the latter do not have the energy density when viewed individually, so that macro-bodies (quantum astrophysics II) always remain unaffected due to contravariant general laws of nature between these parallel worlds.

(10) F Supergravity = F Quantum gravity = F Quantum gravity law I – F Quantum gravity law II = $G_1(K_0 - K_1) - G_2(K_0 + K_1) = G \times (Tm\,/\,r \times 1/V)^2 = (1\,/\,\sqrt{(1 - (1/V)^2)}) - 1 = 0$

$\sqrt{-g} = 1$.

Side note to (10).

(10.1) $G \times [(T_1m_1 \times T_2m_2)\,/\,r^2] \times (L/V^2 \times v^2/2) - [G \times [(T_1m_1 \times T_2m_2)\,/\,r^2] \times (-L/V^2 \times v^2/2)] = L\,[(1\,/\,\sqrt{(1 - (v/V)^2)}) - 1]$

Kolek, Erik (2024). On the physical foundations of interstellar space travel. In: *Chronicles of Business Informatics Physics (CBIP)*. Volume 2, edition no. 1.0. ISBN: 9783759731159.

(10.2) $G \times [(T_1 m_1 \times T_2 m_2) / r^2] \times (L/V^2 \times v^2/2) + G \times [(T_1 m_1 \times T_2 m_2) / r^2] \times (L/V^2 \times v^2/2) = L [(1 / \sqrt{(1 - (v/V)^2)}) - 1]$

(10.3) $2[G \times [(T_1 m_1 \times T_2 m_2) / r^2] \times (L/V^2 \times v^2/2)] = L [(1 / \sqrt{(1 - (v/V)^2)}) - 1]$

(10.4) $2[G \times (T^2 m^2 / r^2) \times (L/V^2 \times v^2/2)] = L [(1 / \sqrt{(1 - (v/V)^2)}) - 1]$

(10.5) $2[G \times (Tm / r)^2 \times (L/V^2 \times v^2/2)] = L [(1 / \sqrt{(1 - (v/V)^2)}) - 1]$

(10.6) $2[G \times (Tm / r)^2 \times (1/V^2 \times v^2/2)] = (1 / \sqrt{(1 - (v/V)^2)} - 1$

(10.7) $G \times (Tm / r)^2 \times 1/V^2 \times v^2 = (1 / \sqrt{(1 - (v/V)^2)} - 1$

(10.8) $G \times (Tm / r \times 1/V \times v)^2 = (1 / \sqrt{(1 - (v/V)^2)} - 1$

(10.9) $G \times (Tm / r \times 1/V)^2 = (1 / \sqrt{(1 - (1/V)^2)} - 1$ ///// $V = C$

(10.10) $G \times (Tm / r \times 1/C)^2 = (1 / \sqrt{(1 - (1/C)^2)} - 1 = 0$

Kolek, Erik (2024). On the physical foundations of interstellar space travel. In: *Chronicles of Business Informatics Physics (CBIP)*. Volume 2, edition no. 1.0. ISBN: 9783759731159.

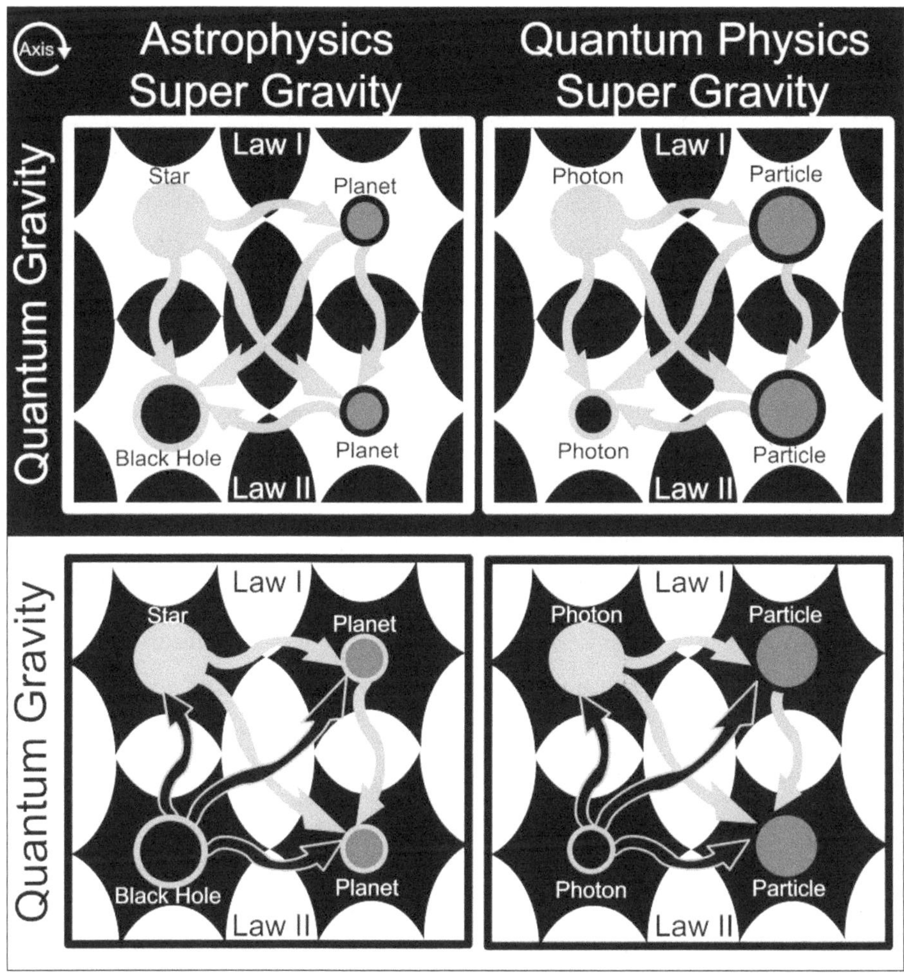

Figure 3. Supergravity fields derived using an advanced theory of the heat of light radiation.

In conclusion, black holes cannot have gravity G (directly) at their event horizon in our expanding universe based on (a), (b), (c) and (d).

(a) Every black hole appears to be growing or expanding in our apparently expanding universe, but they can also catapult out matter and can therefore also look hairy as Stephen Hawking (2016) had assumed. The black holes in our apparently expanding universe (A) therefore most likely represent tornadoes (Figure 1).

Kolek, Erik (2024). On the physical foundations of interstellar space travel. In: *Chronicles of Business Informatics Physics (CBIP)*. Volume 2, edition no. 1.0. ISBN: 9783759731159.

(b) Every black hole has a supergravitational field inside, which means that a hollow ellipsoidal body could exist inside and a gravitational field could also exist outside according to the general theory of relativity (Einstein, 1916), only directly at the event horizon, where all the matter probably collects and moves like dark black and red cloud wind formations, there can be no gravitation G and no gravitational field (Figure 2 and Figure 3). This could also be the reason why a black hole with a few solar masses could appear to astronomers as a shadow due to the gravitational field but with background noise due to radiation.

(c) A further crosscheck for (a) and (b) leading to (d) could apply: The Ferent theory of gravity (2016) does not correspond to real physics but to formal mathematics, since it cannot be applied to a black hole as a very large body and its event horizon according to invariance theory. The reason for this is that no time could be conceivable within a black hole due to supergravity, at least no linear homogeneous time that would have to exist as a generally contravariant time with respect to the special theory of relativity (Einstein, 1905), but directly at the event horizon of a black hole no gravity and thus no time can be assumed to exist. The latter is also independent of whether it forms a vortex or tornado with quantum matter or not. A lack of time in this real physics does not represent a conceptual limitation of the quantum theory of gravity.

(d) The gravity G may be removed from the Hawking temperature (6), (7) and (8) (Hawking, 1974; Hawking, 1975) with respect to the event horizon of a black hole on the basis of (a), (b) and (c). Otherwise, the Hawking temperature (Hawking, 1974; Hawking, 1975) would not agree with the assumed dark universe model (Figure 2). The completed Hawking temperature (8) (Hawking, 1974; Hawking, 1975) appears to be true not only for black hole tornadoes possibly existing in our apparently expanding universe, but it is also conceivably true with respect to black hole swirls that could possibly exist in another (according to this principle parallel) shrinking universe.

Kolek, Erik (2024). On the physical foundations of interstellar space travel. In: *Chronicles of Business Informatics Physics (CBIP)*. Volume 2, edition no. 1.0. ISBN: 9783759731159.

As a final consequence, I would like to say that quantum gravitational fields including both laws according to the general expression for their power rotation potential $\Sigma\int dx_1 dx_2 dx_3 dx_4 dp_1 dp_2 dp_3 dp_4$ lead to the physical determination of a likewise rotating supergravitational field represented within our apparently expanding universe, which can be added as two new additional forces to the standard model of (present-day) physics. An as yet undetermined type of mega-gravitational field could also be conceivable as a connecting consequence outside our apparently expanding universe, but would not necessarily be relevant for an updating or extension of the standard model of physics, since new insights could be expected from this, but these would of course have to be mentally equivalent to a meta-quantum physics based only on abstract optics-dynamics. Moreover, the knowledge regarding a megagravitational field is not necessarily groundbreaking for the understanding of living beings, for example if a reference body is to move faster than the heat of light radiation c in terms of time, because the knowledge of the laws of quantum gravity and supergravity theory should already be satisfactory for this without a metamodel. The Hawking temperature (Hawking, 1974; Hawking, 1975) therefore remains physically justified, which means that Ferent (2016) is wrong with his assumption in this respect. Ferent (2016) is again wrong with his statement that Albert Einstein's (1905; 1916) special and general theory of relativity would be limited with regard to the speed of light c, since this is merely an interchangeable condition within the coincident covariant equation systems and with regard to the heat of light, Ferent (2016) does not find any expression. Such an imagined limitation in the form of the speed of light or light radiation heat is (for me) no barrier to recognize and describe the supergravity theory including the quantum gravity laws. At the same time, and in this one moment, this new unified or superlative gravitational field theory, supported by the findings of Albert Einstein (1905; 1916) and Stephen Hawking (1974; 1975), probably represents the most suitable principle-related physics to describe how very small bodies such as an electron mass field, but also very large bodies such as an electron mass field, each with "energy" corresponding to the star power, would have to experience an extremely

Kolek, Erik (2024). On the physical foundations of interstellar space travel. In: *Chronicles of Business Informatics Physics (CBIP)*. Volume 2, edition no. 1.0. ISBN: 9783759731159.

fast direction of acceleration dx_{ik} within a practically rigid coordinate system K relative to the moving coordinate system K'. It must be taken into account that such an artificially generated supergravity field must of course be able to push any mass gravitational field of any celestial object (such as planets and stars) out of its direction of motion. In short: For safety reasons (initially without any further practical research experience), supergravity fields may only be artificially created outside the (respective) own solar system in order to make interstellar voyages of discovery tangible for mankind for the first time.

In the meantime, while I was writing this last episode, another photo of a black hole was published by the observers from Earth [this photo can be viewed on the following website: https://eventhorizontelescope.org/blog/astronomers-image-magnetic-fields-edge-m87s-black-hole (last visited on 07.06.2024)]. It shows the observer an electromagnetic field that looks the same as a black hole vortex or black hole tornado from above, regardless of its current position. This physical reality observed from the Earth represents excellent evidence to confirm the described supergravity field theory given by the two quantum gravity laws coincident with general laws of nature (initially only) covariant for our space-time continuum as a whole by means of theory but not yet confirmed by discovery experience.

References

Einstein, A. (1905). Ist die Trägheit eines Körpers von seinem Energiegehalt abhängig? *Annalen der Physik 18(13)*, pp. 639–641.

Einstein, A. (1916). Die Grundlage der allgemeinen Relativitätstheorie. *Annalen der Physik 354(7)*, pp. 769–822.

Ferent, A. (2016). Ferent Gravitation theory. [https://www.academia.edu/23507496/Ferent_Gravitation_theory (visited on 08.02.2021]).

Hawking, S. W. (1974). Black hole explosions?. Nature. 248 (5443): pp. 30–31.

Kolek, Erik (2024). On the physical foundations of interstellar space travel. In: *Chronicles of Business Informatics Physics (CBIP)*. Volume 2, edition no. 1.0. ISBN: 9783759731159.

Hawking, S. W. (1975). Particle creation by black holes. Communications in Mathematical Physics. 43 (3): pp. 199–220.

Hawking, S. W., Perry, M. J., and Strominger, A. (2016). Soft Hair on Black Holes. Phys. Rev. Lett. 116, No. 23, 231301.

Newton, I. (1687). *Philosophiae Naturalis Principia Mathematica.* 1. edition. Jussu Societatis Regiae ac typis Josephi Streater, London 1687 (http://cudl.lib.cam.ac.uk/view/PR-ADV-B-00039-00001/9 [visited on 08.02.2021]).

Schwarzschild, K. (1916). Über das Gravitationsfeld eines Massenpunktes nach der Einsteinschen Theorie. Sitzungsberichte der Königlich Preussischen Akademie der Wissenschaften 7, pp. 189–196.

Table of contents

In this research article, an advanced theory of black holes is developed. A distinction was made between tornadoes and black hole vortices. In the end, it was the tornadoes of black holes that could probably exist in the physical world. All black holes most likely have no gravity G at their event horizons.

Kolek, Erik (2024). On the physical foundations of interstellar space travel. In: *Chronicles of Business Informatics Physics (CBIP)*. Volume 2, edition no. 1.0. ISBN: 9783759731159.

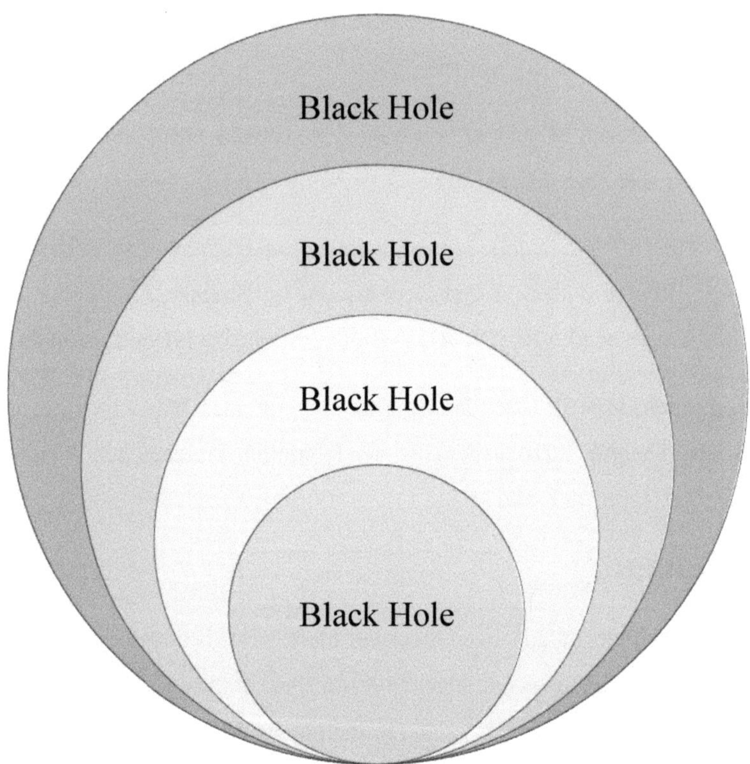

Figure 4. Table of contents illustrated by the importance of the advanced theory of black holes.

Kolek, Erik (2024). On the physical foundations of interstellar space travel. In: *Chronicles of Business Informatics Physics (CBIP)*. Volume 2, edition no. 1.0. ISBN: 9783759731159.

Scientific citation:

Kolek, Erik (2024). On a heuristic viewpoint regarding a quantum field theory for the generation and conversion of heat of light as an advanced molecular kinetic theory of light and heat. In: *On the physical foundations of interstellar space travel.* Chronicles of Business Informatics Physics (CBIP). Volume 2, edition no. 1.0.

Erik Kolek (2024)

On a heuristic viewpoint regarding a quantum field theory for the generation and conversion of heat of light as an advanced molecular kinetic theory of light and heat

Summary

This theoretical treatise deals with the study of quantum radiation and superradiation emanating from stars and black holes. A new assumption regarding the human perception of light and shadow emerges, thanks to the introduction of the concepts of light and dark photons. Overall, this is a light and dark theory of relativity based on blackbody radiation.

On a heuristic viewpoint regarding a quantum field theory for the generation and conversion of heat of light as an advanced molecular kinetic theory of light and heat

Hypothesis

The results of the radiation investigation published by Albert Einstein in his annus mirabilis in 1905 lead to a very inspiring assumption, which is now to be applied to light (Figure 1). This work is related to original work by Albert Einstein and Stephen William Hawking. To my knowledge, their physical realities are the only working foundations for an advanced molecular kinetic theory of light and heat.

Kolek, Erik (2024). On the physical foundations of interstellar space travel. In: *Chronicles of Business Informatics Physics (CBIP)*. Volume 2, edition no. 1.0. ISBN: 9783759731159.

According to Einstein (1905a), "[...] classical thermodynamics can no longer be regarded as strictly valid even for microscopically distinguishable spaces [...]" (p. 549). I think I could be the researcher who could decide the important research question on the theory of heat (Einstein, 1905c). Therefore, I need to study the quantum radiation of stars and black holes to learn more about light and heat.

Since the mass of the photons (light) is hot, especially at the beginning of the radiation, I say that the energy L is equal to the temperature T measured in Kelvin. This equivalence principle is covered by equation (1). If a star, as an example of a very large body, emits the temperature T (energy L) in the form of mass and radiation, its mass decreases by T/V^2 and the hot mass of photons (light) is accelerated positively by $v^2/2$. The kinetic energy of a star is therefore given by equation (1). V stands for the speed of the light in the form of photons and v for the speed of their hot mass in the form of quantum matter. It is therefore understandable that the kinetic energy K_0 in relation to the system (x', y', z', t') differs from the kinetic energy K_1 in relation to another system (x, y, z, t). This difference $K_0 - K_1$, which results from equation (1), has a multi-dimensional physical logic that is easy to understand. K_0 and K_1 represent the kinetic energy values of an identical body or a star, but are associated with two coordinate systems existing in relative motion. Consequently, a star or a black hole in one of the systems (system (x', y', z', t')) (practically) not in motion and another very large body is in relative motion to this system (system (x, y, z, t)).

(1) $K_0 - K_1 = T/V^2 \times v^2/2 = (+T/V^2 \times v^2/2)$

If a bright photon, as an example of a very small body, emits the temperature T (energy L) in the form of mass and heat (radiation), its mass decreases by T/V^2 and the photon volume (heat of light) is negatively accelerated by $v^2/2$. The kinetic energy of a bright photon is therefore given by equation (2). Again, V stands for the speed of light and heat of the photons and v for the speed of their volume change in the form of hot quantum matter. In this model assumption, the heat also has a mass as a fraction of the photon mass when the mass is viewed with the magnifying glass of quantum

Kolek, Erik (2024). On the physical foundations of interstellar space travel. In: *Chronicles of Business Informatics Physics (CBIP)*. Volume 2, edition no. 1.0. ISBN: 9783759731159.

physics. Imagine our expanding universe if there were only photons at one level of the quantum spectrum.

How would photons transfer heat to each other, measured as temperature T (energy L)? As can be seen in Figure 1, the quantum radiation comes from a star and goes directly into a black hole. It can therefore be assumed that the photons become smaller and smaller on their way into this black hole. While the mass of the photons loses volume, the heat is also lost or released into space-time in the form of temperature T (energy L). When I think of solar rays, I tend to think of solar winds in the form of a hot mass of photons traveling at the speed of light c as twisted solar clouds on their way from a star (or our sun) to a planetary surface (or to Earth). This twisted hot photon cloud (a sunbeam) can be seen as a very bright light by an observer outside the planet's atmosphere and feels hotter the closer the observer gets to the star (or our sun). So photons could cool down so much that they seem to disappear in our expanding universe, but that doesn't seem to be the truth either. Photons might not disappear, they might change from light to dark, sinking deeper into the quantum spectrum of physics.

(2) $K_0 - K_1 = T/V^2 \times (-v^2/2) = (-T/V^2 \times v^2/2)$

On the other hand, if a black hole absorbs the temperature T (energy L) in the form of hot mass and radiation, its mass increases by T/V^2 and the light (hot mass of photons) is also positively accelerated by $v^2/2$. The negative kinetic energy of a black hole is therefore given by equation (3). V again means the speed of light in the form of photons and v again means the speed of their hot mass in the form of quantum matter. Therefore, it is again understandable that the kinetic energy K_0 with respect to the system (x', y', z', t') is different from the kinetic energy K_1 with respect to another system (x, y, z, t), as I have already explained. This difference $K_0 + K_1$, which results from equation (3), naturally has a multi-dimensional physical logic that is easy to understand. K_0 and K_1 represent the kinetic energy values of an identical body or black hole, but are associated with two coordinate systems existing in relative motion. Thus,

Kolek, Erik (2024). On the physical foundations of interstellar space travel. In: *Chronicles of Business Informatics Physics (CBIP)*. Volume 2, edition no. 1.0. ISBN: 9783759731159.

a black hole or a star in one of the systems (system (x', y', z', t')) (practically) not in motion and another very large body is in relative motion to this system (system (x, y, z, t)).

(3) $K_0 + K_1 = (-T/V^2) \times v^2/2 = (-T/V^2 \times v^2/2)$

If a dark photon, as an example of a very small body, absorbs the temperature T (energy L) in the form of mass and heat (radiation), its mass increases by T/V^2 and the photon volume (heat of light) is positively accelerated by $v^2/2$. The kinetic energy of a dark photon is therefore given by equation (4). Again, V stands for the speed of light and heat of the photons and v for the speed of their volume change in the form of hot quantum matter. It should be remembered that in this model assumption the heat also has a mass as a proportion of the photon mass. As a consequence of this equivalence assumption of heat as mass, the dark photons would have to be much smaller than the light photons. Dark photons are probably still invisible to experimental physics today. In our expanding universe, there could therefore be a reaction or interaction between light photons and dark photons, which everyone on planet Earth recognizes as light and heat.

(4) $K_0 + K_1 = (-L/V^2) \times v^2/2 = (-L/V^2 \times v^2/2) = (2) K_0 - K_1 = L/V^2 \times (-v^2/2) = (-L/V^2 \times v^2/2)$

The two laws of quantum radiation (1) and (3) are modelled and visualized supersymmetrically and relativistically in order to provide an overview and demonstrate their significance for astrophysics and quantum physics. This model-based visualization covers all previously clarified physical facts. This visualization represents a model section of our expanding universe. It is intended to make the quantum radiation laws (1) and (3) easy to understand without in-depth mathematical knowledge. Both quantum radiation laws could be applied in astrophysics and quantum physics, because they refer both to the hot mass and radiation of stars and black holes interacting with each other and to the mass and heat (radiation) of light particles and dark particles of quantum matter interacting with each other as light

Kolek, Erik (2024). On the physical foundations of interstellar space travel. In: *Chronicles of Business Informatics Physics (CBIP)*. Volume 2, edition no. 1.0. ISBN: 9783759731159.

photons and dark photons. In an open system, dark matter particles can become positive and not negative like light photons due to the quantum radiation law II (3). In an open system, light particles could therefore be connected to dark photons via heat and the quantum spectrum could be dark despite the quantum radiation law I (1). Both types of particles (light and dark photons) could therefore be treated as anti-particles (anti-photons) and could never be separated from each other. Perhaps a sunbeam with the temperature T (energy L) represents a molecular-kinetic reaction, but also a molecular-electromagnetic reaction between already existing anti-photons. A sunbeam could therefore be a special kind of lightning in which the temperature T (energy L) is used to generate photons at the speed of light V and to transfer heat to each other via the volume change velocity v. In the special and general theory of relativity according to Albert Einstein (1905; 1916), which applies to astrophysics, an identical multidimensional physical logic appears to exist for quantum physics.

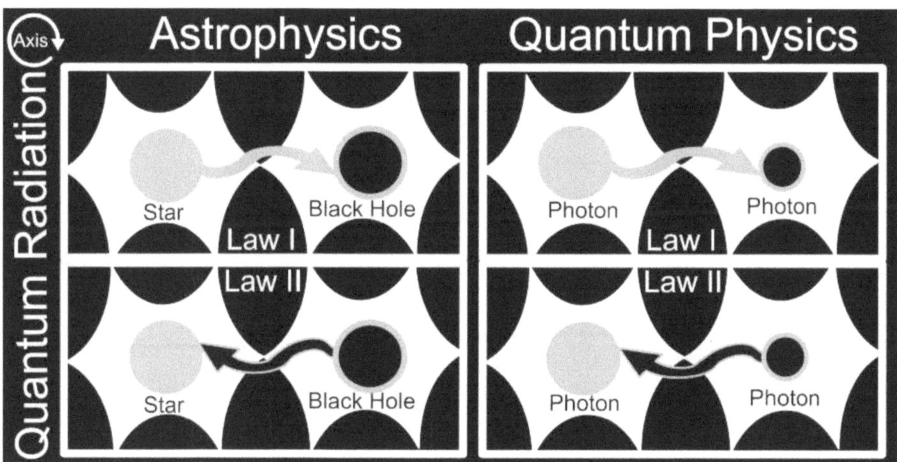

Figure 1. Quantum radiation in astrophysics and quantum physics.

So what could light be? Light could represent a physical reaction of two anti-photons due to temperature T (energy L), which is seen as a flash by an observer with the magnifying glass of quantum physics (Figure 2). This flash could contain a trialism consisting of a wave, particle and energy or a wave, particle and temperature. Light

Kolek, Erik (2024). On the physical foundations of interstellar space travel. In: *Chronicles of Business Informatics Physics (CBIP)*. Volume 2, edition no. 1.0. ISBN: 9783759731159.

photons could merge with dark photons and light and temperature should emerge, which can be seen and felt by an observer on all kinds of (very large) bodies, not only on a planet like Earth. In this idea of the quantum model, the observer could see trillions of very small stars while believing that light is shining. While trillions of very small stars are shining, no very small black hole (dark photon) can emerge from the much deeper quantum spectrum (again). So in this quantum model idea, the observer could also see trillions of very small black holes while believing that the light is not shining (I mean shadows).

Although trillions of very small black holes could exist in the shadow, their quantum gravity should be so extremely low that matter with much larger mass, such as humans in astrophysics, cannot be harmed, since all living things can also live in the shadow. As I said before, the observer could see trillions of very small black holes at night while thinking that no light is shining. Very small black holes could appear right at the moment (point event) when all bright photons including their temperature T (energy L) have disappeared or better been consumed by dark photons. Consequently, light could arise via bright photons interacting with dark photons, which convert the energy L into heat (temperature T), which an observer with the magnifying glass of quantum physics could see as an electromagnetic flash at any moment (point event) as soon as an anti-photon receives the energy L (temperature T). Heat and temperature changes of anti-photons should also be explained by quantum gravity. Both types of photons could exist and move in energetic particle wave lines in our expanding universe. Anti-photons (as a type of anti-particle) could exist as geometric spheres due to quantum gravity and simultaneously as elongated ellipses due to supergravity.

So how could light be connected to heat? Light in the form of two anti-photons (tiny stars interacting with tiny black holes) could be mass, and heat in the form of a mass fraction between these photons. Imagine if there were only anti-photons with different energy or temperature levels on their surfaces, so that heat could only be transferred via an anti-photon volume change after being produced by interacting tiny stars and

Kolek, Erik (2024). On the physical foundations of interstellar space travel. In: *Chronicles of Business Informatics Physics (CBIP)*. Volume 2, edition no. 1.0. ISBN: 9783759731159.

tiny black holes, thus obtaining the energy L (temperature T) to realize light, which is seen as a tiny flash for each anti-photon reaction. Then heat could be transferred via an anti-photon volume change and that seems to be the truth for light photons and dark photons measurable as different energy or temperature levels as given in light and shadow.

But of course it remains the case that anti-photons are very small bodies and could not behave like macro-stars and macro-black holes, but they could behave supersymmetrically and relativistically in parallel. Smaller anti-photons could then be colder than larger anti-photons. Large anti-photons could be hot and small anti-photons could therefore be cold. Large anti-photons could cool down and become smaller and smaller. Small anti-photons could grow when they get hot and so on.

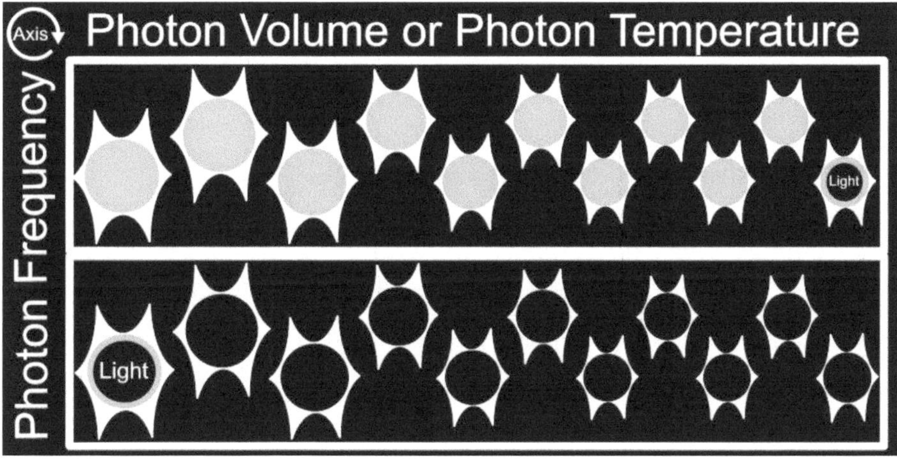

Figure 2. Frequency and volume of light and dark photons could exist in quantum physics.

Evaluation

For a black hole, the quantum spectrum of the emitted and/or absorbed matter is thermal and is given for the emission as a radiation temperature of the black body generally accepted in physics – the Hawking temperature is given in natural units and $T_H = k / 2\pi$ (Hawking, 1974; 1975). In the international system of units (abbreviated

Kolek, Erik (2024). On the physical foundations of interstellar space travel. In: *Chronicles of Business Informatics Physics (CBIP)*. Volume 2, edition no. 1.0. ISBN: 9783759731159.

SI units), it is converted to $T_H = hk / 2\pi c k_B$ (Hawking, 1974; 1975). The Schwarzschild gravity in SI units for the surface of a black hole is characterized by $k = c^4 / 4GM$ (Schwarzschild, 1916). For the mass M of the black hole, the Hawking temperature is then inversely proportional to equation (6). The Hawking temperature thus comprises the reduced Planck constant (h), the light constant (c), the gravitational constant (G), the mass of the black hole (M) and the Boltzmann constant (k_B) (Hawking, 1974; 1975).

(6) $T_{H1} = hc^3 / 8\pi GMk_B$ (mass decrease leads to temperature rise)

Therefore, the Hawking temperature seems to be compatible with the model assumption: If a black hole emits temperature T (energy L) in the form of mass and radiation, its mass decreases by T/V^2. Here the Hawking temperature increases, but it should also increase if a black hole absorbs the temperature T in the form of mass and radiation, its mass increases by T/V^2. I therefore obtain a second Hawking temperature (7).

(7) $T_{H2} = 8\pi GMk_B / hc^3$ (mass increase leads to temperature increase)

For a black hole that absorbs and emits mass and radiation, the full Hawking temperature is given by equation (8). All terms are now based on an advanced molecular kinetic theory of light and heat, which supersymmetrically explains the temperature changes at the event horizon during the ongoing mass adaptation of a black hole.

(8) $T_{H3} = hc^3 / 8\pi GMk_B + 8\pi GMk_B / hc^3$ (mass change leads to temperature change)

Since in quantum physics the energy L (temperature T) of an anti-photon could also be a type of energy in the form of gravity G, I must replace G, which is relevant for astrophysics, with L, which is relevant for quantum physics. For a black hole that absorbs and releases mass and radiation, the complete Hawking temperature including the energy L instead of the gravity G is given by equation (9).

Kolek, Erik (2024). On the physical foundations of interstellar space travel. In: *Chronicles of Business Informatics Physics (CBIP)*. Volume 2, edition no. 1.0. ISBN: 9783759731159.

(9) $T_{H4} = hc^3 / 8\pi LMk_B + 8\pi LMk_B / hc^3$ (mass change leads to energy change)

If the energy L is zero, the temperature T should also be zero, which means that at this point in time (point event) there is no temperature phenomenon at the event horizon of a (tiny) black hole. If the energy L increases, the temperature T should also increase as a function of the mass of a (tiny) black hole. Thus I obtain the new Einstein-Hawking energy L_{EH1} (10) for a stationary anti-photon after transforming equation (9).

(10) $L_{EH1} = T / (hc^3 + 8\pi Mk_B)$ (mass decrease leads to temperature rise)

(10.1) $T = hc^3 / 8\pi LMk_B + 8\pi LMk_B / hc^3 \; // \times hc^3$

(10.2) $T \times hc^3 = h^2c^6 / 8\pi LMk_B + 8\pi LMk_B \; // \times 8\pi LMk_B$

(10.3) $T \times hc^3 \times 8\pi LMk_B = h^2c^6 + (8\pi LMk_B)^2 \; // \; /(hc^3 \times 8\pi LMk_B)$

(10.4) $T = hc^3 + 8\pi LMk_B \; // \; /L$

(10.5) $T / L = hc^3 + 8\pi Mk_B \; // \; /T$

(10.6) $1 / L = (hc^3 + 8\pi Mk_B) / T$

(10.7) $L / 1 = T / (hc^3 + 8\pi Mk_B)$

Starting from the Einstein-Hawking energy L_{EH1} (10), I can also define the Einstein-Hawking mass M_{EH1} (11).

(11) $M_{EH1} = (T - hc^3) / 8\pi k_B$

(11.1) $L \times (hc^3 + 8\pi Mk_B) = T$

(11.2) $Lhc^3 + 8\pi LMk_B = T$

(11.3) $8\pi LMk_B = T - Lhc^3$

(11.4) $M = (T - Lhc^3) / 8\pi Lk_B$

Kolek, Erik (2024). On the physical foundations of interstellar space travel. In: *Chronicles of Business Informatics Physics (CBIP)*. Volume 2, edition no. 1.0. ISBN: 9783759731159.

(11.5) $M = (T - hc^3) / 8\pi k_B$

I have also succeeded in finding an advanced Kolek-Boltzmann constant (k_{KB}) (12) on the basis of the Einstein-Hawking mass M_{EH1} (11), which should improve all conceivable equations that still contain the previous Boltzmann constant (k_B). An advanced constant therefore generally leads to advanced physics.

(12) $k_{KB} = (T - hc^3) / 8\pi M$

(12.1) $8\pi M k_B = (T - hc^3)$

(12.2) $k_B = (T - hc^3) / 8\pi M$

For a black hole that absorbs and releases mass and radiation, the full Hawking temperature based on energy L is given by equation (9), which is used to find an advanced Kolek-Planckian quantum of action (h_{KP}) (13).

(13) $h_{KP} = T_{H4} / (c^3 + 8\pi LM k_B)$

(13.1) $T_{H4} = hc^3 / 8\pi LM k_B + 8\pi LM k_B / hc^3$

(13.2) $T_{H4} \times hc^3 = h^2 c^6 / 8\pi LM k_B + 8\pi LM k_B$

(13.3) $T_{H4} \times hc^3 \times 8\pi LM k_B = h^2 c^6 + (8\pi LM k_B)^2$

(13.4) $T_{H4} \times h = h^2 c^3 + 8\pi LM k_B$

(13.5) $1 / h = (c^3 + 8\pi LM k_B) / T_{H4}$

(13.6) $h / 1 = T_{H4} / (c^3 + 8\pi LM k_B)$

For a black hole that absorbs and emits mass and radiation, the full Hawking temperature based on the energy L is given by equation (9), which can also be used to determine an extended Kolek heat constant of light (c_K) (14). The Kolek heat constant of light (c_K) explains how quickly heat is transferred via light not only in space-time

Kolek, Erik (2024). On the physical foundations of interstellar space travel. In: *Chronicles of Business Informatics Physics (CBIP)*. Volume 2, edition no. 1.0. ISBN: 9783759731159.

but also in liquids at rest. This Kolek light heat constant (c_K) could enable advanced laser technology, known as phaser technology.

(14) $c_K = 3\sqrt{[T_{H4} / (h + 8\pi LMk_B)]}$

(14.1) $T_{H4} = hc^3 / 8\pi LMk_B + 8\pi LMk_B / hc^3$

(14.2) $T_{H4} \times hc^3 = h^2c^6 / 8\pi LMk_B + 8\pi LMk_B$

(14.3) $T_{H4} \times hc^3 \times 8\pi LMk_B = h^2c^6 + (8\pi LMk_B)^2$

(14.4) $1 / c^3 = (h + 8\pi LMk_B) / T_{H4}$

(14.5) $c^3 / 1 = [T_{H4} / (h + 8\pi LMk_B)]$

(14.6) $c = 3\sqrt{[T_{H4} / (h + 8\pi LMk_B)]}$

Since the complete Hawking temperature (9) contains the energy L instead of the gravitation G due to the equivalence principle, some new findings about the gravitation G and the energy L are also possible due to the equations (10), (11), (12), (13) and (14), because I also get some new Einstein-Hawking gravitational equations (15a), (16a), (17a), (18a) and (19a), and some new Einstein-Hawking energy equations (15b), (16b), (17b), (18b) and (19b); but for the sake of mathematical completeness, there are also simplified statements (15c), (16c), (17c), (18c), and (19c) with various dependencies possible via the complete Hawking temperature (9). These simplified statements came about because I repeatedly made mathematical approximations. All equations represent consequences of a new Einstein-Hawking theory, which is a research work beyond the standard model of physics, and thus a new theory of everything, which is necessary to successfully extend this standard model of physics. To further improve this new theory of everything, I (I think) need to travel to a black hole and observe its behavior in space-time, for example with a warp spaceship.

(10) $L_{EH1} = T_{H4} / (hc^3 + 8\pi Mk_B)$

Kolek, Erik (2024). On the physical foundations of interstellar space travel. In: *Chronicles of Business Informatics Physics (CBIP)*. Volume 2, edition no. 1.0. ISBN: 9783759731159.

(11) $M_{EH1} = (T_{H4} - hc^3) / 8\pi k_B$

(12) $k_{KB} = (T_{H4} - hc^3) / 8\pi M$

(13) $h_{KP} = T_{H4} / (c^3 + 8\pi LMk_B)$

(14) $c_K = 3\sqrt{[T_{H4} / (h + 8\pi LMk_B)]}$

(15.a) $G_{EH1} = T_{H4} / (hc^3 + 8\pi Mk_B)$

(15.b) $L_{EH1} = G / (hc^3 + 8\pi Mk_B)$

(15.c) $T_{EH1} = G_{EH1} \times (hc^3 + 8\pi Mk_B)$

(16.a) $M_{EH1} = (G - hc^3) / 8\pi k_B$

(16.b) $M_{EH1} = (L - hc^3) / 8\pi k_B$

(16.c) $M_{EH1} = (T - hc^3) / 8\pi k_B$

(16.d) $G_{EH1} = 8\pi Mk_B + hc^3$

(16.e) $L_{EH1} = 8\pi Mk_B + hc^3$

(16.f) $T_{EH1} = 8\pi Mk_B + hc^3$

(17.a) $k_{KB} = (G - hc^3) / 8\pi M$

(17.b) $k_{KB} = (L - hc^3) / 8\pi M$

(17.c) $k_{KB} = (T - hc^3) / 8\pi M$

(17.d) $G_{EH1} = 8\pi Mk_{KB} + hc^3$

(17.e) $L_{EH1} = 8\pi Mk_{KB} + hc^3$

(17.f) $T_{EH1} = 8\pi Mk_{KB} + hc^3$

(18.a) $h_{KP} = G / (c^3 + 8\pi LMk_B)$

Kolek, Erik (2024). On the physical foundations of interstellar space travel. In: *Chronicles of Business Informatics Physics (CBIP)*. Volume 2, edition no. 1.0. ISBN: 9783759731159.

(18.b) $h_{KP} = L / (c^3 + 8\pi GMk_B)$

(18.c) $h_{KP} = T / (c^3 + 8\pi GMk_B)$

(18.d) $G = h_{KP} \times (c^3 + 8\pi LMk_B)$

(18.e) $L = h_{KP} \times (c^3 + 8\pi LMk_B)$

(18.f) $T = h_{KP} \times (c^3 + 8\pi LMk_B)$

(19.a) $c_K = 3\sqrt{[G / (h + 8\pi LMk_B)]}$

(19.b) $c_K = 3\sqrt{[G / (h + 8\pi TMk_B)]}$

(19.c) $c_K = 3\sqrt{[L / (h + 8\pi GMk_B)]}$

(19.d) $c_K = 3\sqrt{[L / (h + 8\pi TMk_B)]}$

(19.e) $c_K = 3\sqrt{[T / (h + 8\pi GMk_B)]}$

(19.f) $c_K = 3\sqrt{[T / (h + 8\pi LMk_B)]}$

(19.g) $c_K = 3\sqrt{[1 / (h + 8\pi Mk_B)]}$

(19.h) $G_{EH1} = c_K^3 \times (h + 8\pi LMk_B)$

(19.i) $G_{EH1} = c_K^3 \times (h + 8\pi TMk_B)$

(19.j) $L_{EH1} = c_K^3 \times (h + 8\pi GMk_B)$

(19.k) $L_{EH1} = c_K^3 \times (h + 8\pi TMk_B)$

(19.l) $T_{EH1} = c_K^3 \times (h + 8\pi GMk_B)$

(19.m) $T_{EH1} = c_K^3 \times (h + 8\pi LMk_B)$

(19.n) $1 = c_K^3 \times (h + 8\pi Mk_B)$

(19.o) $h_{EH1} = 1 / c_K^3 - 8\pi Mk_B$

Kolek, Erik (2024). On the physical foundations of interstellar space travel. In: *Chronicles of Business Informatics Physics (CBIP)*. Volume 2, edition no. 1.0. ISBN: 9783759731159.

(19.p) $k_{EH1} = (1 / c_K^3 - h_{EH1}) / 8\pi M$

(19.q) $M_{EH1} = (1 / c_K^3 - h_{EH1}) / 8\pi k_B$

(19.r) $\pi_{EH1} = (1 / c_K^3 - h_{EH1}) / 8M k_B$

Equation 20 could correspond to the primordial force of the sun from which the universe was once born in the big bang; it is therefore likely that humanity could find itself in a black hole, which in turn could also ignite back into a star (at any time by itself) through friction with another universe. Another destructive big bang by means of which our black universe (black hole) could probably become a bright universe (star) again, in which no living beings can exist due to the high Hawking temperature. This Einstein-Hawking theory probably actually corresponds to a physically real theory of everything, which describes the possible gravitational collapse.

(20) $L = E = Tmc^2 = Gmc^2$

If I imagine a circle at the equator of a sphere or rather an ellipse, surprisingly, some new insights about the size of our expanding universe also emerge. Perhaps the geometry equations (21), (22), (23) and (24) could help to design and build a metal spherical frame around the sun or other planets and stars, because a warp spaceship could also be designed and built now that mankind has a new definition of pi (π). Doctors can write π with or without the expansion velocity v. Based on our observations, I know in general that our universe expands positively at the speed v, so I assume that it does not shrink negatively and therefore all equation results must also be positive, and in particular this physical fact is important for the universe equation (23). In this case, (23) π forms the best estimator for the size of the universe because most of the constants generally accepted in physics are included, except for the mass M, which we should have already approximated for visible and invisible quantum matter. The question for experimental physics is now: How big is the universe in reality and what can mankind learn from this for the development of space technology?

Kolek, Erik (2024). On the physical foundations of interstellar space travel. In: *Chronicles of Business Informatics Physics (CBIP)*. Volume 2, edition no. 1.0. ISBN: 9783759731159.

This new pi can be used in any geometric equation containing π, especially in (theoretical) physics, but also in other disciplines such as mathematics.

(21) $v\pi = (Ghc^3 - h^2c^6) / 8LMk_B$

(22) $v\pi = (T_{H4}hc^3 - h^2c^6) / 8GMk_B$

(23) $-v\pi = (-hc^3) / 8Mk_B = v\pi = (hc^3) / 8Mk_B$

(24) $v\pi = (T_{H4}hc^3 - h^2c^6) / 8LMk_B$

(24.1) $T_{H4} = hc^3 / 8\pi LMk_B + 8\pi LMk_B / hc^3$

(24.2) $T_{H4} \times hc^3 \times 8\pi LMk_B = h^2c^6 + (8\pi LMk_B)^2$

(24.3) $T_{H4}hc^3 - h^2c^6 = 8\pi LMk_B$

(24.4) $\pi = (T_{H4}hc^3 - h_2c^6) / 8LMk_B$

Since light and heat could be a reaction of anti-photons that could be electromagnetically concatenated and therefore only the energy L (temperature T) in a coordinate system (x', y', z', t') is in kinetic motion relative to the coordinate systems (x, y, z, t) of the anti-photons, I can assume that the temperature T of an anti-photon on its spherical or more likely elliptical surface should behave like the temperature T of a black hole at its event horizon.

So if a moving anti-photon, as an example of a very small body, releases the Einstein-Hawking energy L in the form of mass and heat (radiation), its mass decreases by L_{EH1}/V^2 and its surface (light) is accelerated by $v^2/2$. This also gives me the relevant equation (25) for moving stars in astrophysics.

(25) $K_0 - K_1 = L_{EH1}/V^2 \times v^2/2 = ([T / (2 + 8\pi Mk_B)]/V^2 \times v^2/2)$

On the other hand, if a moving anti-photon absorbs the Einstein-Hawking energy L in the form of mass and heat (radiation), its mass increases by L_{EH1}/V^2 and its surface

Kolek, Erik (2024). On the physical foundations of interstellar space travel. In: *Chronicles of Business Informatics Physics (CBIP)*. Volume 2, edition no. 1.0. ISBN: 9783759731159.

(light) is also accelerated by $v^2/2$. Therefore, I also get the relevant equation (26) for moving very large bodies such as black holes in astrophysics.

(26) $K_0 + K_1 = -(L_{EH1}/V^2 \times v^2/2) = -([T / (2 + 8\pi Mk_B)]/V^2 \times v^2/2)$

By transforming the formula (10), I obtain the Einstein-Hawking temperature T for a stationary anti-photon (27).

(27) $T_{EH1} = L_{EH1} \times (2 + 8\pi Mk_B)$ (mass decrease leads to energy decrease)

Based on Einstein (1906) for Brownian motion in the case of a moving anti-photon, which floats in a space-time with $t = 1$ and moves along the x-axis, I obtain V_x (28) for the speed of motion of this very small body and V_r (29) for its angular velocity. P is the radius of a sphere, R is the gas constant, N is the number of anti-photons in one gram of spacetime.

(28) $V_x = \sqrt{(R[L_{EH1} \times (2 + 8\pi Mk_B)]/N \times 1/3\pi k_B P)} = \sqrt{(R[L_{EH1} \times (2 + 8M)]/ 3NP)}$

(29) $V_r = \sqrt{(R[L_{EH1} \times (2 + 8\pi Mk_B)]/N \times 1/4\pi k_B P^3)} = \sqrt{(R[L_{EH1} \times (1 + 4M)]/ 2NP^3)}$

Since it will be very difficult in experimental quantum physics to count all the anti-photons in spacetime, I recommend dropping R and N from the photon motion (28) and (29), which leads to a more down-to-earth theory of velocity (30) and (31). For the quantum motion of an anti-photon, its rotational velocity V_r as a function of its radius P appears to be almost equal to its velocity of travel V_x on the x-axis. An increase in P slows down the rotational speed V_r of the anti-photon and slightly less its velocity V_x. The situation is completely different with the mass of a photon. An increase in M accelerates the anti-photon's speed of travel V_x by twice as much and its speed of rotation V_r by only half as much.

(30) $V_x = \sqrt{([L_{EH1} \times (2 + 8M)]/ 3P)}$

(31) $V_r = \sqrt{([L_{EH1} \times (1 + 4M)]/ 2P^3)}$

Kolek, Erik (2024). On the physical foundations of interstellar space travel. In: *Chronicles of Business Informatics Physics (CBIP)*. Volume 2, edition no. 1.0. ISBN: 9783759731159.

In general, it must be remembered that I have exchanged the gravitational force G for the energy L for the motion of a very small body, because all relevant quantum field theories in quantum physics should be unified. By converting (30) and (31), I obtain the second Einstein-Hawking energy L_{EH2}, which is necessary to move an anti-photon on the x-axis (32) and to control its rotational speed, both depending on its mass (33).

(32) $L_{EH2} = (V_x^2 \times P) / (3/2 + 3/8M)$

(33) $L_{EH2} = (V_r^2 \times P^3) / (1/2 + 2M)$

For astrophysics, I assume from (33) that the gravity G of a moving body or sphere is also dependent on its rotational speed (33) while it is moving on the x-axis (32). As soon as bodies are accelerated, this dependency should increase. So if I have on one side of the x-axis the gravity G of a black hole, as an example of a curved singularity, and on the other side a planet, like the Earth, which rolls like a soccer in this deep singularity and becomes heavier in the process, both speeds should also increase. Again, I assume and apply the equivalence principle between gravity and energy, because both can have the same effect, namely acceleration. Thus I obtain L_{EH2} from the equation of the Einstein-Hawking energy with quantum gravity (34).

(34) $(V_x^2 \times P) / (3/2 + 3/8M) + (V_r^2 \times P^3) / (1/2 + 2M) = G \times [(m_1 \times m_2) / r^2] \times -([T / (2 + 8\pi Mk_B)]/V^2 \times v^2/2)$

In order to describe the acceleration effect of gravity G or energy L, a dependency due to the interaction between masses, distances, velocities, light and heat, expressed as temperature, could exist as an explanation. Therefore, the gravitational force G can be exchanged with the Einstein-Hawking energy L_{EH1} (35).

(35) $(V_x^2 \times P) / (3/2 + 3/8M) + (V_r^2 \times P^3) / (1/2 + 2M) = [T / (2 + 8\pi Mk_B)] \times [(m_1 \times m_2) / r^2] \times -([T / (2 + 8\pi Mk_B)]/V^2 \times v^2/2)$

So if heat also has a mass and if heat represents a mass fraction of a body, gravity should mainly depend on distances and velocities. So if a very large body such as a

Kolek, Erik (2024). On the physical foundations of interstellar space travel. In: *Chronicles of Business Informatics Physics (CBIP)*. Volume 2, edition no. 1.0. ISBN: 9783759731159.

star has a higher temperature than a star with the same mass, it should still have a higher gravitational force G and energy L. For planets, the closer they are to a star, the greater their temperature-mass ratio should have an increasing effect on their gravity.

Newton's theory of gravitation (1687) could be improved in the sense of higher accuracy if I insert temperature and relative velocity V_{rel}, which an observer has as seen from an unmoving body with progression and rotation velocity V_0, whereby both bodies including their masses are in relative motion to each other in two coordinate systems. This results in a Newton-Einstein theory of gravity (36).

(36) $V_{rel} = V_{x0} - V_{x1} + V_{r0} - V_{r1}$ AND $[G] \times (T_1 m_1 \times T_2 m_2) / (r V_{rel})^2$

The gravity G in (36) can be replaced by the Einstein-Hawking energy L_{EH1} (10), which leads to (37), because of the equivalence principle between the gravity G and the energy L and because both can have the same effect, namely acceleration. For example, when a sovereign spaceship consumes energy to fly, or when it moves because the gravity of a very large body acts on it, both are perceived as acceleration by an observer inside.

(37) $V_{rel} = V_{x0} - V_{x1} + V_{r0} - V_{r1}$ AND $[T / (2 + 8\pi M k_B)] \times (T_1 m_1 \times T_2 m_2) / (r V_{rel})^2$

Stokes' rule is (38) $v_2 \leq v_1$ and means that a subsequent light quantum or photon should contain the same or less energy than the preceding light quantum or photon in the matter wave (Einstein, 1905b). According to Einstein (1905b), the energy L should therefore always be constant or it could decrease along its path, which consists of a photon-matter wave, but it will never be zero because light quanta or photons should always glow, regardless of how large or small they are. Outside of a very large body in spacetime, like in a star or black hole, I can assume that the temperature of $v_2 \leq v_1$ as well as its energy L $v_2 \leq v_1$.

(38) $v_2 \leq v_1$ (Stokes' rule I: Energy L of $v_2 \leq v_1$ if the temperature of $v_2 \leq v_1$)

Kolek, Erik (2024). On the physical foundations of interstellar space travel. In: *Chronicles of Business Informatics Physics (CBIP)*. Volume 2, edition no. 1.0. ISBN: 9783759731159.

For Einstein (1905b), two deviations from Stokes' rule are possible: (a) the amount of energy produced is so large that a light quantum or photon can gain energy from several light quanta or photons, and (b) if the light produced is not equal in its energy L, as is the case with black radiation according to Wien's law, for example if the light is produced by a star or black hole with extremely high temperatures, where Wien's law no longer applies to the wavelength.

I agree with Einstein (1905b) that a non-Vienna radiation, even in large reduction, could behave in a different energy ratio than a blackbody radiation as a valid form of Vienna radiation. Consequently, I assume a second Stokes' rule (39) $v_2 \geq v_1$ for a non-Vienna radiation. In a very large body in spacetime, such as a star or black hole, the temperature of v_2 would therefore have to be $\geq v_1$, and the energy L of v_2 would also have to be $\geq v_1$. This is also due to the fact that if v_2 is closer to the center of such a very large body than v_1, it should also be hotter, not only because of the pressure, but also because more photons interact or merge into each other the closer it gets to the center of the star or black hole.

(39) $v_2 \geq v_1$ (Stokes' rule II: Energy L of $v_2 \geq v_1$ if the temperature of $v_2 \geq v_1$)

If I think of black radiation emitted from a black hole, (38) Stokes' rule I should also apply to radiation outside a black hole. Outside a black hole, dark photons should be pulled apart from all dimensions due to gravity and the heat decreases as the photons move away from the center. Inside a black hole, the radiation could be different from this physical point of view, (39) Stokes' rule II should apply to the radiation inside a black hole. Inside a black hole, the dark photons should be compressed due to gravity and the heat increases the closer the photons get to the center.

At least some of the black or dark radiation should never escape from a black hole, but it is constantly trying to escape and some of it is pulled back every time it tries because that kind of radiation would also have to be matter. When I think of a star, I would imagine this is also true for light radiation, but to a much lesser extent because

Kolek, Erik (2024). On the physical foundations of interstellar space travel. In: *Chronicles of Business Informatics Physics (CBIP)*. Volume 2, edition no. 1.0. ISBN: 9783759731159.

of the lower gravity compared to a black hole. This makes me think differently about light (and heat).

So if light is only visible when there is an interaction due to the energy L of light and dark radiation or photons, it should be clear when it is light (day) or dark (night). In our expanding universe, it would always be dark everywhere if no bright radiation is emitted by stars. Without stars, black holes would have to be responsible for the darkness everywhere. Perhaps black holes even look different in an expanding universe with more than four dimensions. Then a black hole would look to an observer like a hollow star with a black pearl inside. On the other hand, stars without black holes would have to provide brightness everywhere. Since there are stars and black holes in our expanding universe, dark and light radiation should interact in the form of light, which is represented by the energy L, and a visible matter should change from dark to light. Every time light and dark photons meet, light should be produced as long as the energy L is not completely consumed by dark photons. In my conclusion, dark photons should represent quantum black holes and light photons should represent quantum light stars.

I can take this even further and therefore assume that a quantum star and a macro star as well as a quantum black hole and a macro black hole should behave similarly and the difference between them should be the heat. If I imagine a micro-black hole with the same mass as a micro-star, the gravity G should still be different because of the heat T. This difference in gravity G should also apply to macro-objects such as stars and black holes, which have the same mass but different heat. I don't know yet if a black hole can be hotter than a star in reality, but that could be the case if there is a star nearby and if that star emits bright photons that are absorbed by that black body. So if I have two very small or very large bodies with the same mass and the heat of one body is higher, the gravity G should also be higher, at least for a moment.

In astrophysics, heat is measured in temperature, and in quantum physics, heat is also measured in temperature, but this heat represents matter and is therefore quantum in

Kolek, Erik (2024). On the physical foundations of interstellar space travel. In: *Chronicles of Business Informatics Physics (CBIP)*. Volume 2, edition no. 1.0. ISBN: 9783759731159.

size. Therefore, macro bodies such as humans feel comfortable heat from quantum stars absorbed by quantum black holes around them while it is light (day). Consequently, macro bodies can resist quantum gravity because they should have higher gravity, or as I say, macro bodies cause super gravity or super radiation.

However, superradiation can also be explained by an advanced light and heat theory (Figure 3) according to Einstein (1905), since c does not change during the emission and absorption of light and heat. Therefore, I use (40) as a fundamental basis for a theory improvement that leads to a unified field theory of radiation.

(40) $K_0 - K_1 = L\,[(1 / \sqrt{(1 - (v/V)^2)}) - 1]$

(41) $K_0 - K_1 = ([T / (2 + 8\pi Mk_B)]/V^2 \times v^2/2)[(1 / \sqrt{(1 - (v/V)^2)}) - 1]$

Since I have to imagine the two coordinate systems in relative motion and the speed of light as a constant c, I can now assume, following Einstein (1916), that everything is quantum matter (including electromagnetic fields and thermal fields), except for the radiation field, which would have to exist as superradiation both in our expanding universe and in a black hole. Thus, I obtain a new Einstein field equation that represents a unified radiation field theory based on the speed of light, which explains two types of light – light and dark light, represented by colliding light and dark photons.

(42) F Superradiation = F Quantum radiation = F Quantum radiation law I – F Quantum radiation law II = $G_1(K_0 - K_1) - G_2(K_0 + K_1) = 2/V^2 \times v = 1 / \sqrt{(1 - (1/V)^2)} - 1 = 0$

$\sqrt{-g} = 1.$

(42.1) $G_1(K_0 - K_1) - G_2(K_0 + K_1) = (L/V^2 \times v^2/2) + (L/V^2 \times v^2/2) = L\,[(1 / \sqrt{(1 - (v/V)^2)}) - 1]$

(42.2) $G_1(K_0 - K_1) - G_2(K_0 + K_1) = 2(L/V^2 \times v^2/2) = L\,[(1 / \sqrt{(1 - (v/V)^2)}) - 1] / L$

Kolek, Erik (2024). On the physical foundations of interstellar space travel. In: *Chronicles of Business Informatics Physics (CBIP)*. Volume 2, edition no. 1.0. ISBN: 9783759731159.

(42.3) $G_1(K_0 - K_1) - G_2(K_0 + K_1) = 2(1/V^2 \times v^2/2) = 1 \, [(1 / \sqrt{(1 - (v/V)^2)}) - 1] / v$

(42.4) $G_1(K_0 - K_1) - G_2(K_0 + K_1) = 2(1/V^2 \times v/2) = 1 / \sqrt{(1 - (1/V)^2)} - 1$

(42.5) $G_1(K_0 - K_1) - G_2(K_0 + K_1) = 2/V^2 \times v = 1 / \sqrt{(1 - (1/V)^2)} - 1$

(44) F Superradiation = F Quantum radiation = F Quantum radiation law I – F Quantum radiation law II $= G_1(K_0 - K_1) - G_2(K_0 + K_1) = 2 = 1 / \sqrt{(1 - (v/V)^2)} - 1$

$\sqrt{-g} = 1.$

(44.1) $G_1(K_0 - K_1) - G_2(K_0 + K_1) = (T / (2 + 8\pi Mk_B)]/V^2 \times v^2/2) + (T / (2 + 8\pi Mk_B)]/V^2 \times v^2/2) = (T / (2 + 8\pi Mk_B)]/V^2 \times v^2/2)[(1 / \sqrt{(1 - (v/V)^2)}) - 1]$

(44.2) $2(T / (2 + 8\pi Mk_B)]/V^2 \times v^2/2) = (T / (2 + 8\pi Mk_B)]/V^2 \times v^2/2)[(1 / \sqrt{(1 - (v/V)^2)}) - 1]$

(44.3) $2 = 1 / \sqrt{(1 - (v/V)^2)} - 1$

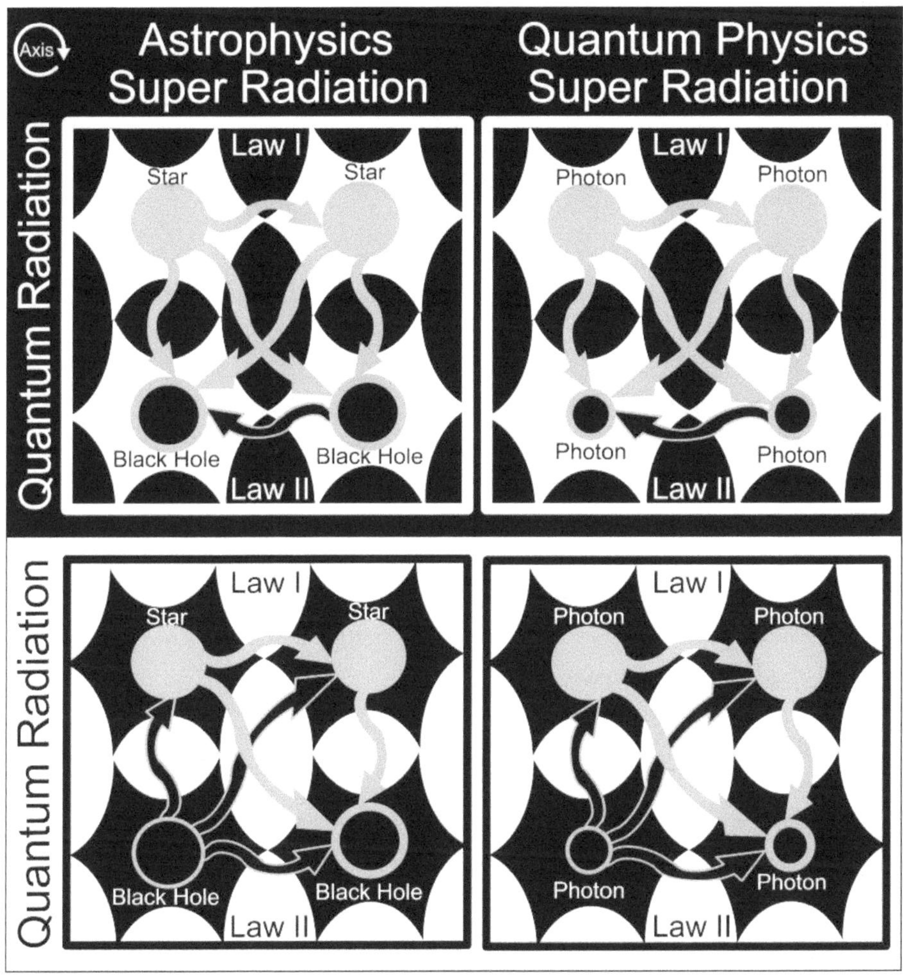

Figure 3. Super radiation derived with an advanced molecular kinetic theory of light and heat.

References

Einstein, A. (1905a). Does the Inertia of a Body Depend upon its Energy Content? *Annalen der Physik* 18, pp. 639–641.

Einstein, A. (1905b). On a Heuristic Point of View Concerning the Production and Transformation of Light. *Annalen der Physik* 17, pp. 132–148.

Kolek, Erik (2024). On the physical foundations of interstellar space travel. In: *Chronicles of Business Informatics Physics (CBIP)*. Volume 2, edition no. 1.0. ISBN: 9783759731159.

Einstein, A. (1905c). On the Movement of Small Particles Suspended in Stationary Liquids Required by the Molecular-Kinetic Theory of Heat. *Annalen der Physik* 17, pp. 549–560.

Einstein, A. (1906). On the Theory of Brownian Motion. *Annalen der Physik* 19, pp. 371–381.

Einstein, A. (1916). Die Grundlage der allgemeinen Relativitätstheorie. *Annalen der Physik* 354(7), pp. 769–822.

Hawking, S. W. (1974). Black hole explosions?. *Nature*. 248 (5443): pp. 30–31.

Hawking, S. W. (1975). Particle creation by black holes. *Communications in Mathematical Physics*. 43 (3): pp. 199–220.

Schwarzschild, K. (1916). Über das Gravitationsfeld eines Massenpunktes nach der Einsteinschen Theorie. *Sitzungsberichte der Königlich Preussischen Akademie der Wissenschaften* 7, pp. 189–196.

Appendix

Since in quantum physics the energy L (gravity G) of an antiphoton could also be a type of energy in the form of the temperature T, I must replace the L relevant to quantum physics with the G relevant to astrophysics. For a black hole that absorbs and releases mass and radiation, the full Hawking temperature including gravity G is given by equation (9a) instead of the energy L.

(9a) $T_{H3} = hc^3 / 8\pi GMk_B + 8\pi GMk_B / hc^3$ (mass change leads to temperature change)

If the gravitational force G is zero, the temperature T should also be zero, which means there is no temperature phenomenon. If gravity G increases, the temperature T should also increase as a function of the mass. This gives me the new Einstein-Hawking gravity G (10a) for a stationary anti-photon.

(10a) $G_{EH1} = T / (2 + 8\pi Mk_B)$ (mass reduction leads to gravity increase)

Kolek, Erik (2024). On the physical foundations of interstellar space travel. In: *Chronicles of Business Informatics Physics (CBIP)*. Volume 2, edition no. 1.0. ISBN: 9783759731159.

(10.1a) $T = hc^3 / 8\pi GMk_B + 8\pi GMk_B / hc^3$ // $\times hc^3$

(10.2a) $T \times hc^3 = 2hc^3 / 8\pi GMk_B + 8\pi GMk_B$ // $\times 8\pi GMk_B$

(10.3a) $T \times hc^3 \times 8\pi GMk_B = 2hc^3 + (8\pi GMk_B)^2$ // $/(hc^3 \times 8\pi GMk_B)$

(10.4a) $T = 2 + 8\pi GMk_B$ // $/G$

(10.5a) $T / G = 2 + 8\pi Mk_B$ // $/T$

(10.6a) $1 / G = (2 + 8\pi Mk_B) / T = G / 1 = T / (2 + 8\pi Mk_B)$

Table of contents

In this research article, an advanced theory of light and heat is developed. It was assumed that there must be light and dark photons. This interaction can be analogized to stars and black holes. Depending on the proximity to the center of the very large body, effects such as increased fusion (light) or cooling (dark) occur.

Kolek, Erik (2024). On the physical foundations of interstellar space travel. In: *Chronicles of Business Informatics Physics (CBIP)*. Volume 2, edition no. 1.0. ISBN: 9783759731159.

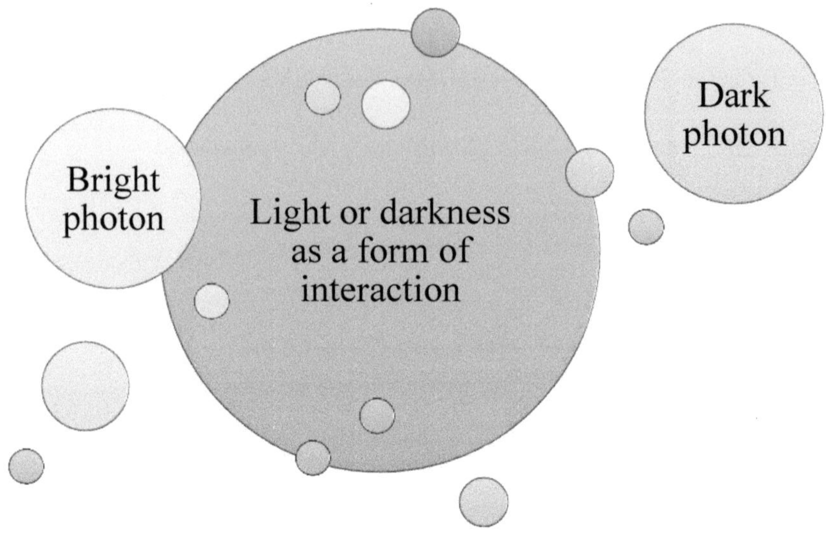

Figure 4. Table of contents illustrated by the importance of the advanced theory of light and heat.